WIRELESS PHONES AND HEALTH II: State of the Science

Erratum: On page 67, 12 lines from the bottom of the page, the equations "$R=\frac{1}{2}D^2/\lambda$ or $R=\frac{1}{4}D^2/\lambda$" should be replaced by "$R=D^2/\lambda$ or $R=2D^2/\lambda$".

WIRELESS PHONES AND HEALTH II

State of the Science

WIRELESS PHONES AND HEALTH II

State of the Science

Edited by

George L. Carlo
Wireless Technology Research, LLC

Associate Editor

Polly M. Thibodeau

Foreword by
George L. Carlo

KLUWER ACADEMIC PUBLISHERS
Boston / Dordrecht / London

Distributors for North, Central and South America:
Kluwer Academic Publishers
101 Philip Drive
Assinippi Park
Norwell, Massachusetts 02061 USA
Telephone (781) 871-6600
Fax (781) 681-9045
E-Mail <kluwer@wkap.com>

Distributors for all other countries:
Kluwer Academic Publishers Group
Distribution Centre
Post Office Box 322
3300 AH Dordrecht, THE NETHERLANDS
Telephone 31 78 6392 392
Fax 31 78 6546 474
E-Mail <services@wkap.nl>

Electronic Services <http://www.wkap.nl>

Library of Congress Cataloging-in-Publication Data

Wireless phones and health II: state of the science / edited by George L. Carlo; associate editor, Polly M. Thibodeau; foreword by George L. Carlo.
p. cm.
Includes bibliographical references and index.
ISBN 978-0-7923-7977-5 (alk. paper)
1. Telephones, Wireless—Health aspects—Congresses. 2. Radio waves—Health Aspects—Congresses. I. Carlo, George Louis. II. Thibodeau, Polly.

RA569.3 .W552 2000
612'.01448—dc21

00-058421

Printed on acid-free paper.

The Publisher offers discounts on this book for course use and bulk purchases. For further information, send email to <joanne.tracy@wkap.com> .

Contents

Section V: Epidemiology

Section VI: Future Directions

Appendices

Second State the Science Colloquium
on the Public Health Impact of Wireless Technology
Long Beach, California
19-20 June 1999

Foreword

From the time questions about the impact of wireless technology on public health were first raised in 1993 through the present, Wireless Technology Research, LLC (WTR) has been the largest independent surveillance and research program trying to identify and solve human health problems associated with wireless phones. In 1995 at the University "La Sapienza" of Rome, WTR sponsored the first comprehensive forum for the discussion of these issues. Papers from the 1995 State of the Science Colloquium were collected and published in Volume I of this series, *Wireless Phones and Health: Scientific Progress*. This second volume assembles papers presented at WTR's Second State of the Science Colloquium in Long Beach, CA, in June 1999; it contains the most comprehensive research on the public health impact of wireless phones to date.

The operating words for the proper understanding of these data are science and public health. Science is a tool for making public health decisions, but the framework in which we are operating is truly that of public health. We are looking for problems that have to do with wireless technology. We are trying to decide how this technology impacts on the public for one purpose and one purpose alone, and that purpose is to solve problems that are identified.

I would like to challenge you, the reader, to suspend your parochial orientation as you consider these latest findings. While it is important to know who has funded research and while it is important to know what perspective the researcher might be bringing, good science speaks for itself.

I would like to make a final point about the problem that we are addressing. Wireless technology is the most rapidly deploying technology in the history of mankind. We have never had a technology or a consumer product permeate the world as fast and as efficiently as wireless technology has. At the end of 1999 there were about 90 million wireless phone users in the United States alone. I have heard estimates of over 500 million users of wireless technology worldwide currently. With such a pervasive technology, it is incumbent upon us to make sure that as this technology evolves, we maintain our ability to understand what the technology means in terms of public health. If the pervasion of the technology outstrips our ability to explain it in terms of public health, we are not doing our job. I would therefore like to underscore the importance of this volume by saying that there are 500 million people around the world who are counting on us to get this right.

G.L. Carlo

Acknowledgments

The papers in this volume represent the current state of the science in the study of potential health effects from wireless telecommunication technology.

I would like to thank all of the scientists who presented papers at the Second State of the Science Colloquium (Long Beach, CA, June 1999) for their contributions to this volume and for their contribution to the increasing body of knowledge on this topic. Additionally, I would like to thank those individuals who chaired the Colloquium sessions, as well as those who attended for their participation in this important scientific exchange.

The Colloquium was held the weekend prior to the 21st Annual Bioelectromagnetics Society meeting, and I owe a great debt to Dr. William Wisecup and the whole of the Bioelectromagnetics Society for coordinating many of the logistical details of our Colloquium.

I would like to thank the staff of Wireless Technology Research, LLC (WTR) for organizing the Colloquium, particularly Nancy Akers. Other WTR staff providing support for the Colloquium included Marjan Najafi, Rebecca Steffens Jenrow, Claudine M. Valmonte, Salem Fisseha, Janis Lemke, Polly M. Thibodeau, Cynthia Perno, Lisa Joson, Bryan W. Eddins, Slavko Bradic, and Robert Bailey.

I would like to thank Polly M. Thibodeau for editing, formatting, and proofreading this volume. The following individuals aided in assembling papers for this volume: Polly M. Thibodeau, Nancy Akers, Marjan Najafi, Claudine M. Valmonte, Salem Fisseha, and Rebecca Steffens Jenrow.

I would also like to thank Mary Panarelli, Carol Cahill, and Joanne Tracy at Kluwer Academic Publishers for their patience and assistance in this process.

The wireless industry, through their sponsorship of WTR, provided financial support for the 1999 Colloquium and this publication.

G.L. Carlo

A portion of the funding required for publication of this volume has been provided by the Bio-Electro-Magnetics Institute of Reno, NV, and by Stetzer Electric, Inc., of Blair, WI.

The Bio-Electro-Magnetics Institute fosters research, education, service, and technical support in matters relating to bioelectromagnetics through studies regarding the relationship between living organisms and exposure to electromagnetic fields. It also studies the health hazards that may result from exposure to sources of radiofrequency radiation, such as radio/TV broadcasts, microwaves, cellular telephones, and radiofrequency fields associated with electric power lines.

Stetzer Electric, Inc., is an electrical contracting firm that has no interest in the wireless telecommunications industry except as a customer using a cellular telephone. Its statement follows: "Support for this publication is part of an effort to educate the public with respect to radiofrequency health hazards associated with advanced technology. Stetzer Electric has an interest in radiofrequency health effects because of the rapidly growing presence of high harmonics in electric power systems, resulting from increased use of electrical apparatus that constitutes a nonlinear load, causing modern electric power to be an increasingly important source of widespread human exposure to radiofrequency fields."

EDITOR'S NOTE: WTR METHODOLOGY

(A detailed overview of WTR is contained in Appendix D.)

HOW WTR WORKS

The ultimate goal of Wireless Technology Research, LLC (WTR) was to identify any public health threats posed by the use of wireless technology early enough to correct them.

The decision-making framework followed by WTR was the public health paradigm with its four operating questions:

1. Is there a public health problem posed by use of wireless instruments?
2. If yes, what is the extent of the problem?
3. What are the appropriate interventions to solve the public health problem?
4. What is the appropriate intervention strategy for those interventions?

Information relevant to public health risk came into the WTR evaluation framework through two pathways: the first was ongoing surveillance whereby WTR scientists evaluated on a continual basis scientific work conducted around the world according to the four operating questions; the second was through WTR's peer-reviewed research program following from the WTR's published Research Agenda.

Prior to making decisions regarding intervention and research, WTR scientific staff sought input from various government, scientific, and industry groups. Industry input was especially important where proprietary technologies were involved and where the only available information was held by the industry.

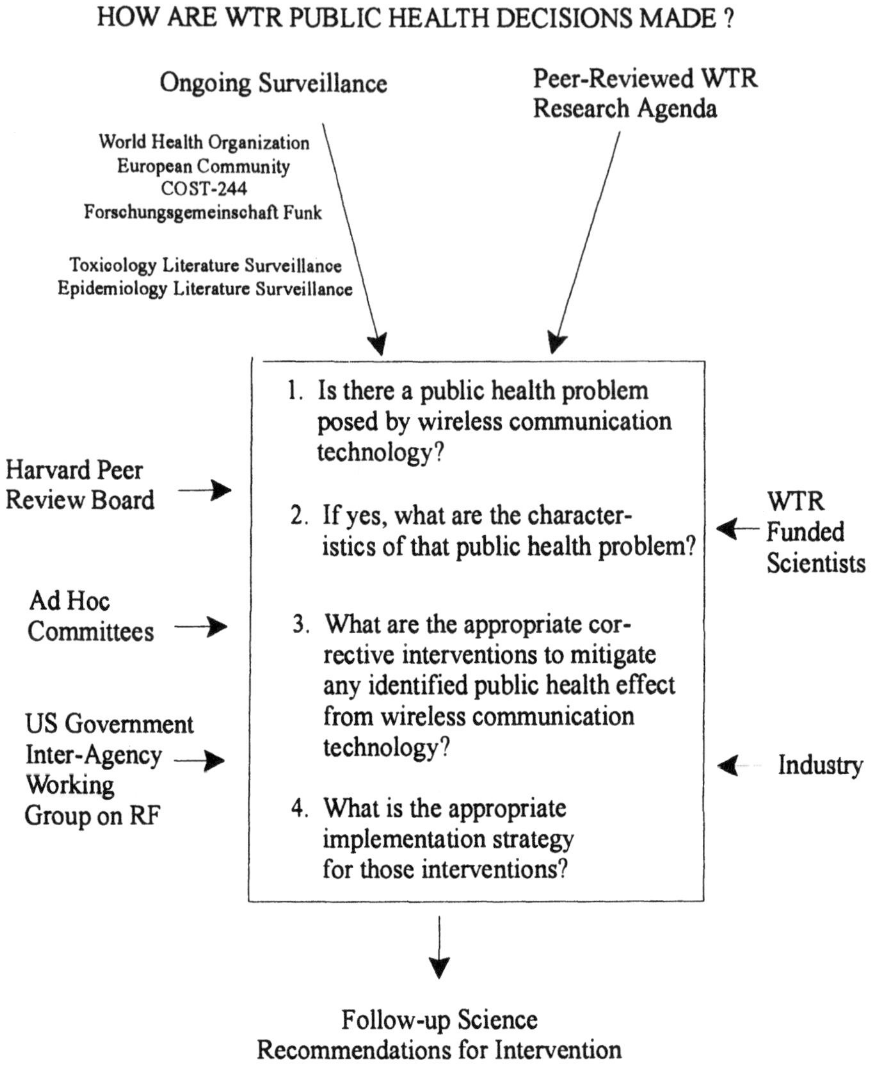

Figure 1. How WTR public health decisions are made.

HOW WTR PUBLIC HEALTH DECISIONS ARE MADE

WTR scientific staff are ultimately responsible for making evaluative decisions within the program. However, it is the policy of WTR to receive input and peer review from a wide range of scientific and government groups (see Figure 1).

The WTR decision-making framework follows the four operating questions of the public health paradigm, and all information potentially useful in identifying public health problems is evaluated against that template.

When a decision about research or intervention is made, WTR initiates appropriate briefings with government and affected industry groups.

WTR TIER I TOXICOLOGY PROGRAM

WTR's Tier I toxicology program was designed to be consistent with standard approaches to pre-market product safety evaluation. While there were no requirements for pre-market approval of wireless phones, the WTR program was designed to build the same database that would be required under a standard pre-market approval protocol.

The WTR program developed new exposure systems for both in vitro and in vivo experiments that allow for assessment of effects in experiments that closely mimic the type of exposures that would be encountered by people using wireless phones. In addition, the WTR exposure systems allowed for assessment of dose-response relationships that are critical to judging the public health significance of toxicological experiments.

WTR TIER II EPIDEMIOLOGY PROGRAM

WTR's Tier II epidemiology program included both cohort and case-control studies conducted in the context of post-market surveillance.

Prior work published by the WTR identified appropriate metrics of measuring exposure to RF radiation through the use of phone company billing records. Those metrics have been employed in the epidemiology research supported by the WTR.

WTR'S SCIENTIFIC DECISION TREE FOR DEFINING PUBLIC HEALTH PROBLEMS

Information was directed into the WTR decision-making framework as a function of either the WTR ongoing surveillance program or its research program. Positive findings in toxicological experiments or positive associations in epidemiological studies were evaluated in terms of public health impact.

The evaluative considerations for public health impact in toxicology and epidemiology were distinct. They involved both objective and subjective assessments of data and required specific expertise for evaluation. Peer review was critical. A positive finding in either a toxicological or a epidemiological study might have lead to a public health intervention, more research, or continued surveillance, but there is no set formula defining what would have been required in a given instance.

The key to making these evaluations in the WTR program was the attempt to make decisions about public health problems on the basis of what is known, rather than on the basis of what is unknown. Within that framework, for example, it was unlikely that a positive finding would rise to a level suggesting public health intervention in the absence of evidence of dose-response, biological plausibility, or concordance between toxicology and epidemiology. However, a positive finding in either toxicology or epidemiology, absent dose-response, would still lead to further research.

The evaluation of scientific data in the context of a public health threat and subsequent decisions about public health interventions could only be made on a case-by-case basis, after thoughtful scientific scrutiny and with rigorous peer review. The WTR decision-making framework encompassed these principles (see Figure 2).

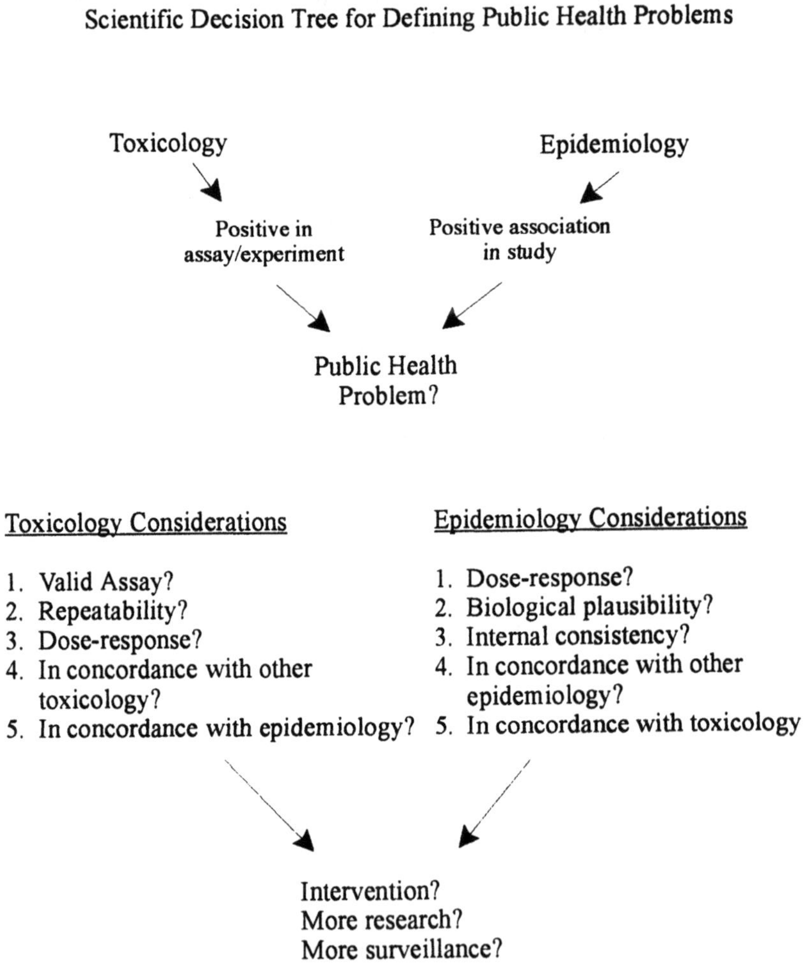

Figure 2. WTR decision-making tree.

I

DOSIMETRY AND MEASUREMENTS

1

LATEST PROGRESS IN DOSIMETRY AND NEAR-FIELD EVALUATIONS FOR MOBILE COMMUNICATIONS DEVICES

N. Kuster

Abstract

The necessity of accurate dosimetric and especially of general near-field evaluations in wireless product development and certification will continuously grow in the coming years. In product development, far-field evaluations provide insufficient information for analyzing and optimizing the performance of transmitters operating in complex environments. Driven by growing public concern, new and more strict legal requirements regarding compliance with safety limits might be released in the near future which will demand enhanced and more precise dosimetric tools and procedures in order to minimize any unnecessary restrictions for the exploration of wireless technology. In addition, special tools will be required to analyze, develop, optimize, and verify exposure systems for bioexperiments addressing public health risk concern. The objective of this paper is to review the latest progress in near-field measurement technology and to discuss future needs.

Introduction

The last decade has seen dramatic development in cellular communications. Consumer desire for always smaller and lighter devices with new features and services coupled with tightened competition has driven manufacturers to cut product cycles to a few months in most cases. In conjunction with this development, radiofrequency (RF) engineers are faced with integrating RF circuits and antennas into ever smaller spaces, yet matching the excellent performance of larger devices with longer antennas, even when operated in very complex environments (such as being held in a hand and pressed against a head). In addition, the devices must also be in compliance with the safety standards for RF exposure (see Table 1).

The future will also see strong development and growth of personal communications networks. The simultaneous use of different systems operating in the near-field of other sources will greatly increase the likelihood of malfunctions due to electromagnetic compatibility (EMC) and electromagnetic interference (EMI) problems. Systems optimized with respect to reliability will be crucial in the case of

	FCC [1]	CENELEC [2]	ARIB STD-T56 [3]
Based on	NCRP [4], [5]	ENV [6]	RCR STD-38 [7]
Group	uncontrolled environment	general public	condition G
Whole-body SAR	0.08 mW/g	0.08 mW/g	0.08 mW/g
Spatial peak SAR	1.6 W/kg	2 W/kg	2 W/kg
Averaging time	30 min	6 min	6 min
Averaging mass	1 g	10 g	10 g
Volume shape	cube	cube	cube
Phantom requirement	not defined	reasonable cross-section of users	several proposed phantoms
Device position	standard	4 positions	normal

Table 1. Basic compliance test requirements proposed in the USA [1], Europe [2], and Japan [3].

life support and enhancement systems. The classical design approach has been to ensure sufficient decoupling of the various elements (electronic devices, RF circuits, matching circuits, antenna, absorbers, reflectors, etc.) either by spatial separation or shielding, such that the different components can be optimized individually. This approach, expensive in cost and space, will not be suitable for the upcoming greatly integrated wireless consumer applications designed to be used in complex environments. Nevertheless, severe unintentional coupling between the components might significantly impair the performance of the device (e.g., the coupling of significant RF currents to other electronic components inside the handset can cause significant energy losses, increased absorption when used next to the head, malfunctions depending on the environment, etc.). As for EMC, it has been sufficient until now to only consider far-field phenomena since operation of devices in close proximity of transmitters has not been an issue. This will change especially with the onset of personal communications applications.

These new applications will require that entire systems be optimized for reliable functioning in various RF environments. In order to enable RF engineers to master this complex task, new sophisticated engineering tools will be needed. In addition to powerful TCAD tools, experimental tools enabling the efficient and accurate analysis of the near-field of prototypes will be essential. The objective of this paper is to review the latest progress in near-field measurement technology and to discuss future needs.

Human Exposure Evaluation

Requirements

Human exposure evaluations become an eminent legal component when performed for demonstrating compliance with safety limits, especially in the context of strong public reservations against electromagnetic exposures. This places another dimension of requirements on the evaluation equipment/procedures with respect to scientific soundness and reliability compared to other similar test requirements (e.g., EMC/EMI, radiation assessment, etc.). There is no doubt that a scientifically sound procedure which can be easily communicated and explicated to experts and non-experts alike would be a great boon. A questionable approach bears the risk of decline in consumer confidence in the mobile communications industry (fewer users and reduced usage) and increased chances for lawsuits, the direct and indirect costs of which could exceed the costs of any evaluation procedure by several magnitudes.

The criteria for a scientifically sound procedure which can be easily communicated are quite obvious and have been outlined in [8]. The standard procedure should

- Ensure that the assessed specific absorption rate (SAR) values do not underestimate the actual maximum exposure occurring in real life situations of a reasonable cross section of the mobile telecommunication equipment (MTE) user group including children;
- Provide reproducible results independent of the testing laboratory; and
- Be comprised of a test procedure whose appropriateness is comprehensible and evident to both experts and non-experts.

The development of instrumentation and procedures satisfying these requirements is a considerable engineering task, since measurements must be conducted inside phantoms with liquids which simulate the dielectric properties of human tissues whereby the source (i.e., the object to be evaluated) is positioned in the closest vicinity of the phantom.

On the other hand, the compliance testing requirements should not unnecessarily obstruct the exploration and advancement of this greatly important technology. This places further requirements on the procedure:

- To avoid being unnecessarily conservative, that is, overestimate the exposure;
- To remain unbiased with respect to the design of the MTE. Low SAR values in the tests should actually be reflected by reduced exposure in real-world situations and vice versa; and
- To allow testing the effects of manufacturing variations.

The assessment of the spatial peak SAR value starting from an unknown field

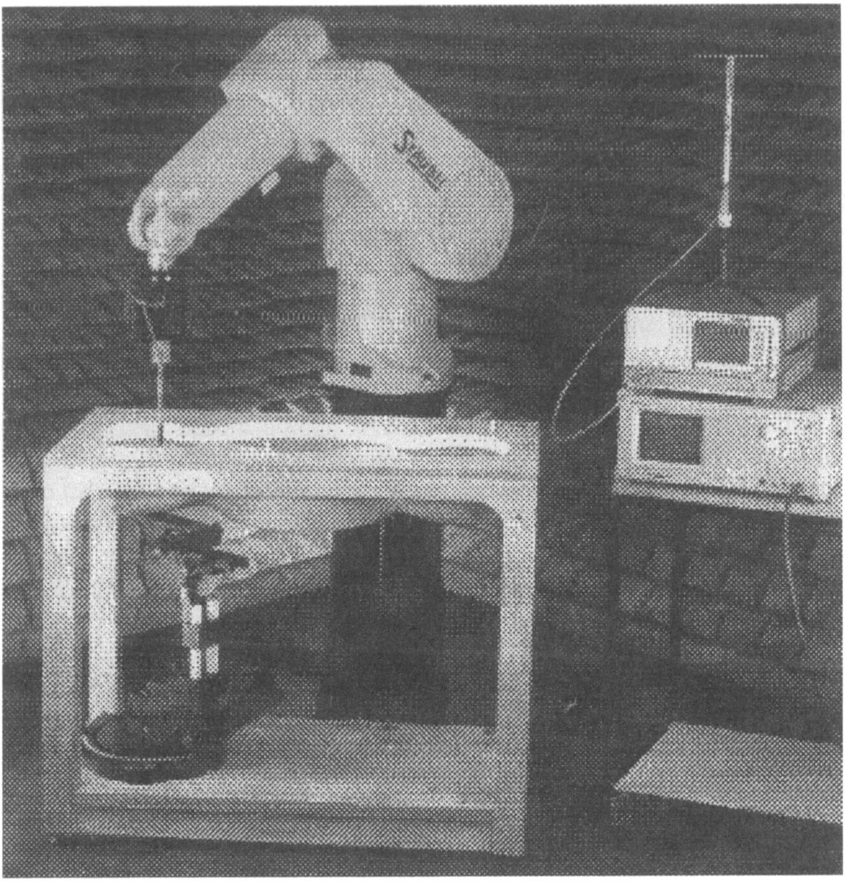

Figure 1. DASY3 near-field measurement system with generic twin phantom.

distribution requires field measurements in several hundred points, data processing, and extrapolation/interpolation between the measured points. Such a task incorporates many different error components, all of which must be thoroughly analyzed in such a manner that is valid for all evaluations. In addition to this class of assessment errors, two other main classes of uncertainties must be evaluated: 1) phantom uncertainty (i.e., uncertainty with respect to the maximum exposure occurring in the user group), and 2) source uncertainty (deviations of the phone tested from those of mass production, manufacturing tolerances, and phone position with respect to the head).

Implementation

A system explicitly designed for the evaluation of human exposure by handheld phones and also satisfying the above criteria is described in [9] and [10]. Figure 1

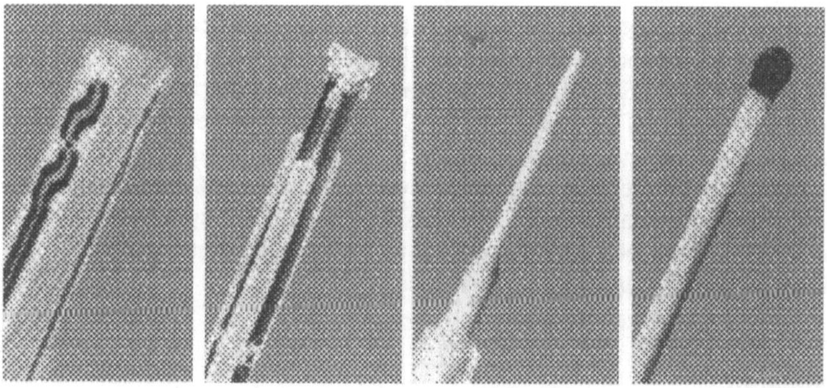

Figure 2. Most recent dosimetric probes. *Left*: probe including a centric optical proximity sensor (outer diameter: 6.8 mm; frequency range: 10 MHz to > 6 GHz; linearity: < ±0.2 dB; full spherical directivity: < ±0.3 dB; dynamic range: 1 μW/g to > 100 mW/g). *Middle left:* compact design with interleaved sensors (outer diameter: 3.9 mm; frequency range: 10 MHz to > 4 GHz; linearity: < ±0.2 dB; full spherical directivity: < ±0.3 dB; dynamic range: 1 μW/g to > 100 mW/g). *Middle right:* 1 mm-resolution probe for special applications such as in vivo and in vitro measurements, characterization of larger probes, and so on. (outer diameter: 1.0 mm; frequency range: 10 MHz to > 10 GHz; linearity: < ±0.2 dB; full spherical directivity: < ±0.2 dB; dynamic range: 20 μW/g to > 200 mW/g). *Right:* a match presented for scale.

shows the third generation of this scanner.

This scanner incorporates a high precision robot (position repeatability < ±0.02 mm), isotropic E-field probes with diode-loaded dipole sensors, an optical proximity sensor for automated positioning of the probe with respect to the phantom surface (precision < ±0.2 mm), and sophisticated software for data processing and measurement control. The generic twin phantom was developed with the aim of covering the maximum exposure that would occur in at least 80% of the entire user population [10]. It enables testing of left-hand, right-hand, and body-mounted usage of devices without requiring a change of phantoms. An integrated test procedure based on a calibrated dipole enables the verification that the system is operating within its specifications.

The key components of the scanner are the probes. In order to achieve the required precision, the probes must be optimized for each particular application with respect to sensitivity, isotropy, linearity, spatial resolution, field distortion (boundary effects, etc.), and immunity (extremely low frequency [ELF] fields, secondary modes of reception, etc.). Field distortions in the vicinity of the probe can result in large errors when the probe is close to boundaries, and field distortion inside the probe can significantly impair its spherical isotropy. Performance tests conducted on several commercially available probes equipped with orthogonally positioned sensors showed deviations from isotropy from ±1.5 up to ±3.4 dB. Figure 2 shows three probes developed for different dosimetric applications, all of which have been optimized to provide excellent isotropy in liquids simulating tissues of high water content (< ±0.3 dB). Probe size reduction generally raises the upper frequency limit, improves spatial resolution, and reduces uncertainty due to boundary effects, but it also results in lower sensitivity and mechanical robustness. The techniques developed for the characterization and optimization of dosimetric probes are described in detail in [11].

A thorough analysis of the uncertainty budget performed in compliance with the NIS81 [12] and the NIST1297 [13] documents has shown that a precision for spatial peak SAR evaluation of better than $< \pm 12$ % (k=1) can be achieved. This is excellent when compared with the common uncertainty inherent in far-field measurements. The uncertainty for the 95% confidence level (k=2) of < 24 % is reasonably low.

In addition to spatial peak SAR assessments, the same system also enables the determination of the total energy absorbed by the user. Such measurements are also relevant for optimizing device performance.

Discussion

The instrumentation described above is an open and flexible implementation of a general near-field scanner enabling the assessment and evaluation of unknown field distributions inside any kind of homogeneous phantom. It is therefore adaptable to the various compliance procedures currently being developed by various organizations (e.g., IEEE, CENELEC, etc.), even though the requirements of these various procedures can be considerably different. For example, the CENELEC approach [2] requires demonstration of compliance for the majority of users (reasonable cross section) and all reasonable ways of using the phone (i.e., most common variations of phone positions with respect to the head). Figure 3 shows test results conducted according to [2]. The large standard deviations between different samples of phones indicate that the uncertainty can only be assessed by evaluating a sufficient number of randomly selected samples.

A more simple approach considered by other standardization bodies is the demonstration of compliance with safety limits induced in the average adult male holding the phone in the position defined by manufacturers, commonly referred to as the intended use position, which in most cases corresponds to the touch position (dot at $\beta = 65°$ and minimum α in Figures 4 and 5). The rationale behind this is that the CENELEC approach is considered to be too restrictive for technology and too time consuming, that is, too many tests are required. This approach also has some disadvantages. In general, tests conducted with a particular head size underestimate the exposure of larger heads. In addition, the touch position for the phone in Figure 4 considerably underestimates the average maximum exposure in real life situations, whereas the same position overestimates the average exposure for the phone in Figure 5.

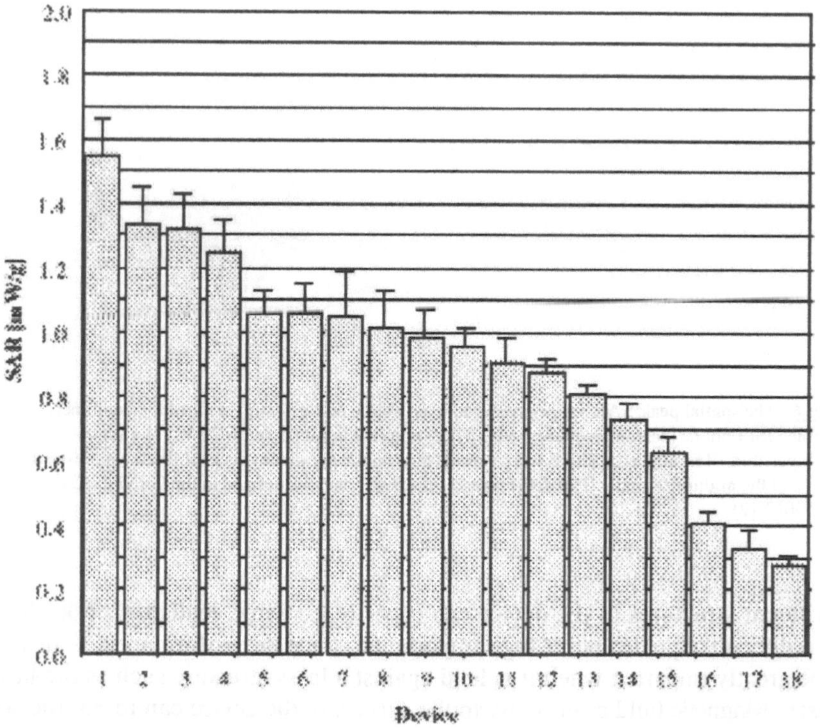

Figure 3. Eighteen GSM devices were tested according to CENELEC prES 59005 [2]. The SAR values shown are the assessed spatial peak SAR values averaged over 10 g and the standard deviation among three or more sample units.

Free Space Evaluations

Requirements

Compliance can be demonstrated not only on the basis of basic limits (i.e., spatial peak SAR limits) but also by showing that the incident field is below the derived exposure limits defined in terms of the electric and magnetic field strengths. This is a suitable approach for indoor base stations, table top and computer-mounted devices, and so on. The requirements regarding instrumentation precision are similar to those of SAR evaluations, whereby extrapolations and volume integration are not required. The boundary effects can be neglected in most cases. On the other hand, such evaluations are more sensitive to reflections in the laboratory environment (e.g., robot, walls, etc.) which must be carefully considered.

An application which is at least as important as compliance testing is the utilization of the scanner for analyzing the RF design of devices. Significant coupling

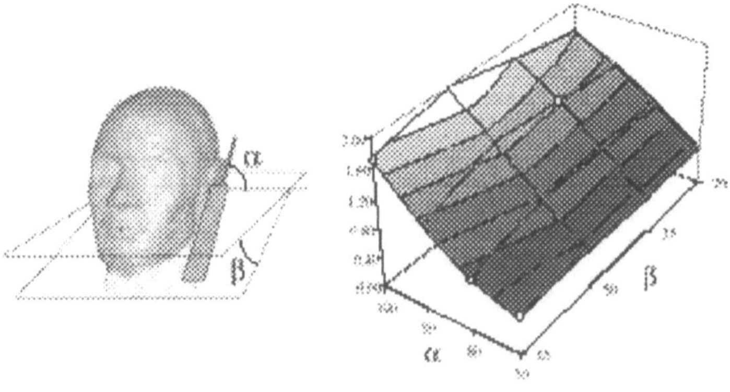

Figure 4. The spatial peak SAR value (averaged over a cube of 1 g tissue) induced by a commercially available GSM phone (carrier frequency: 900 MHz; nominal input power: 0.25 W) as a function of the phone's position. The angles α and β were varied (the phone's ear piece was kept fixed with respect to the entrance of the auditory canal). The dots represent the four positions defined in the CENELEC document prES 59005 [2].

to different structures of the device can occur (e.g., wires, loudspeaker, keyboard). This is difficult to predict but of high importance, since the performance of the device can be greatly impaired when it is held against a lossy structure such as the head of the user. Magnetic field scans close to the surface of the device can reveal the actual current distribution and provide valuable information for evaluating and remedying the source of the problem (see Figure 6). Since the absolute values are of lesser importance, the requirements are lower than in the case of compliance testing. Additional information such as the orientation of the current vector would be of great value. In order to efficiently evaluate printed circuit boards or integrated chips having

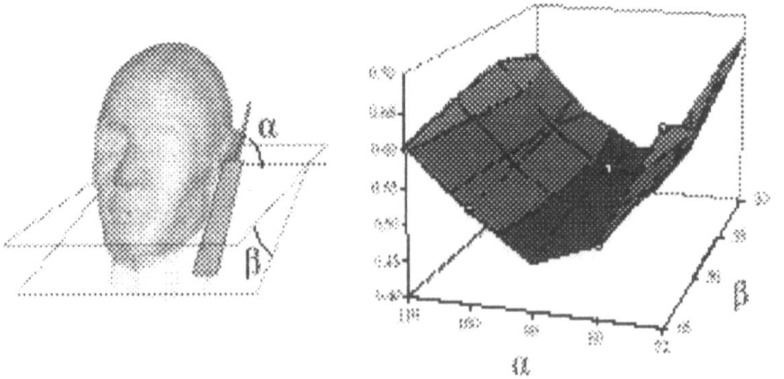

Figure 5. The same evaluation as shown in Figure 4 with a different commercially available GSM phone.

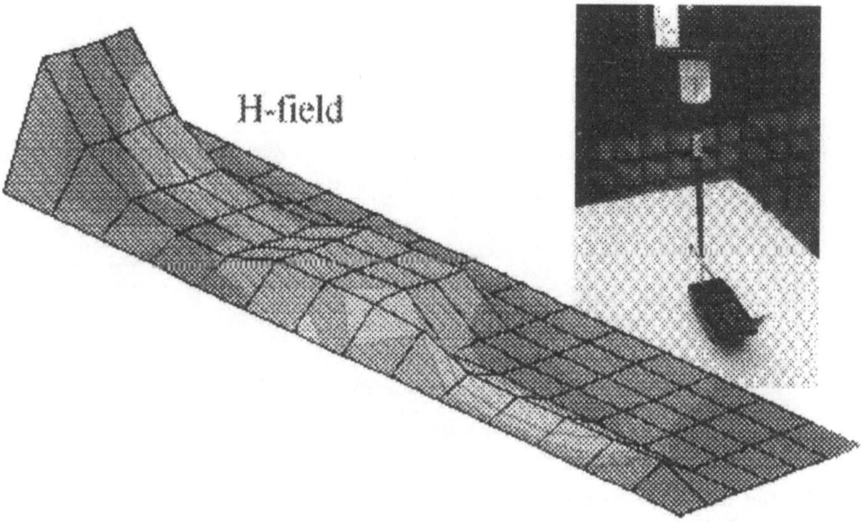

Figure 6. Magnetic field scans of devices provide valuable information to detect RF design deficiencies (e.g., unwanted coupling to loudspeaker, keyboard, battery, wires, etc.).

different frequency components, the complete field information, that is, phase information, is required.

Implementation

The scanner described above can easily be adapted for free-space evaluations provided the appropriate probes are available. Various probes have been developed and optimized for free space measurement and are shown in Figure 7.

Discussion

Detector diode-loaded probes allow only the assessment of the magnitude but cannot provide any phase information about the local field. The local field vector can be assessed by measuring at least six sensor orientations and applying numerical reconstruction of the vector orientations for linear and elliptically polarized fields. A corresponding research project with the objective of developing the corresponding probes and algorithms is close to completion and will be discussed.

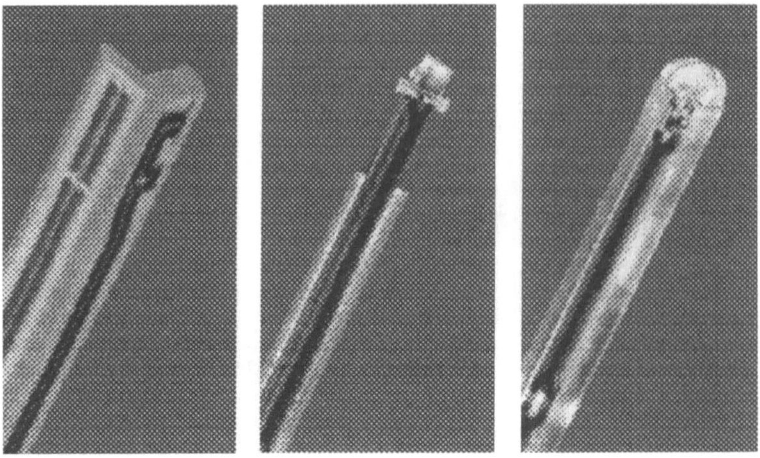

Figure 7. *Left:* free space E-field probe (outer diameter: 8 mm; frequency range: 100 MHz to > 6 GHz; linearity: < ±0.2 dB; directivity: < ±0.4 dB; dynamic range: 2 V/m to > 1000 V/m). *Middle:* free space E-field probe with interleaved sensors (outer diameter: 4 mm; frequency range: 100 MHz to > 6 GHz; linearity: < ±0.2 dB; directivity: < ±0.4 dB; dynamic range: 2 V/m to > 1000 V/m). *Right:* H-field probe (outer diameter: 5 mm; frequency range: 200 MHz to 2.5 GHz; directivity: < ±0.2 dB; dynamic range: 10 mA/m to 2 A/m at 1 GHz).

Radiation Performance Evaluation

Requirements

In most laboratories, the evaluation of the radiation performance of handheld devices is limited to two or three planes of the far-field which do not provide sufficient information about the spherical radiation pattern and the total power radiated. Since the devices should preferably be tested in actual user configurations, the classical planar scanners cannot be used due to a lack of phase stability and synchronization. However, scanners measuring on a spherical surface in the far-field would enable the derivation of the required information. Some products are already available but are expensive in terms of cost and space requirements.

Discussion

An elegant solution would be the near-field scanner configuration as indicated in Figure 8. According to numerical evaluations, radiation pattern and power can be determined with a precision of better than ±0.5 dB for transmitters operating at

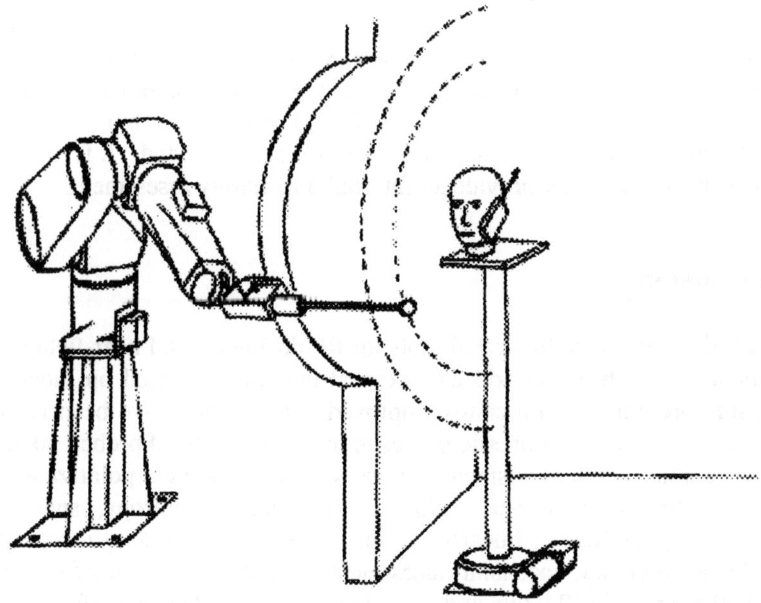

Figure 8. Sketch of a possible implementation of a low-cost scanner for evaluation of the radiation pattern and total radiated power.

frequencies above 800 MHz at an evaluation radius of only 1 m. Although practical implementation will involve additional impediments such as reflections, the development of a prototype is currently in progress.

Experimental Versus Numerical Dosimetry

Among the numerical methods, finite-difference time-domain (FDTD) has become the most powerful simulation technique in the area of dosimetry, favored by the exponential growth in computer performance and continuous improvements of the technique during recent years. The most notable improvements are in the area of absorbing boundary conditions, flexible graded meshes, modeling of media boundaries, source modeling, uncertainty assessments, and graphical user interfaces.

The strengths of experimental dosimetry are in compliance testing and result validation of simulations, since it enables accurate assessment of the induced field strengths within the unmodified/non-simplified real-world setup. The weaknesses include 1) restriction to liquid-like media; 2) limited spatial resolution of larger than 1 mm^3; and 3) limited applicability inside small substructures (i.e., dimensions of the probe tip must be considerably smaller than that of the substructure). The strengths of FDTD are 1) its applicability to handle non-homogeneous structures; 2) provision of high spatial resolution; 3) enabling of complete 3D assessment (e.g., for whole-body or organ-averaged values); and 4) its suitability to evaluate the sensitivity of the results upon different parameters (e.g., dependence of the absorption upon anatomy, posture, tissue parameters, etc.). Therefore, it is well suited for the implementation of TCAD tools for antenna design and EMC analyses. Its most significant weakness is that, in general, real world EMC simulations require simplifications/departures from the actual real-world situation/setup, such that any results must be experimentally validated if error margins smaller than ±2 dB are requested. A further shortcoming is that FDTD does not provide an internal uncertainty assessment.

Conclusion

Recently, several sophisticated tools for RF dosimetry and near-field evaluations of handsets have been introduced. Several new probes based on diode-loaded field sensors providing significantly improved performance have been reported. They range from standard dosimetric probes to highly specialized probes, such as those for in vitro and in vivo assessment, for assessment of the field polarization, and so on. Major advancements were achieved in spherical isotropy, spatial resolution, sensitivity, bandwidth, linearity, immunity against secondary modes of reception, calibration accuracy, uncertainty assessments, and more. The near future will see new tools that are more flexible and provide more comprehensive information about the near-field characteristics of transmitters. Although all these tools are of great use for the evaluation of handsets, the optimization of future applications will require more advanced scanners which accurately provide not only amplitude but also phase information and possess enhanced spatial resolution (in the sub-millimeter range).

Acknowledgments

Some of the cited projects in this paper were generously supported by Motorola, Inc., (US), Ericsson Radio Systems AB (Sweden), NORTEL (Great Britain), T-Mobil GmbH (Germany), Swisscom (Switzerland), and Schmid & Partner Engineering AG (Switzerland), in addition to the Swiss priority research program MINAST and the EUREKA Project SARSYS.

References

[1] FCC. Evaluating compliance with FCC guidelines for human exposure to radiofrequency electromagnetic fields. OET Bulletin 65. Washington, DC: FCC; 1997.

[2] CENELEC. Considerations for evaluation of human exposure to electromagnetic fields (EMFs) from mobile telecommunication equipment (MTE) in the frequency range 30 MHZ - 6 GHz. prES 59005, CLC/TC211 (SEC) 17. Brussels: CENELEC; Mar 1998.

[3] ARIB. Specific absorption rate (SAR) estimation for cellular phone. STD-T56. Jan 1998.

[4] NCRP. Biological effects and exposure criteria for radiofrequency electromagnetic fields.

Report No. 86. NCRP; 1986.

[5] ANSI/IEEE, C95. 1-1992. IEEE standard for safety levels with respect to human exposure to radiofrequency electromagnetic fields, 3 kHz to 300 GHz. New York: ANSI/IEEE; 1992.

[6] CENELEC CLC/TC111B. Human exposure to electromagnetic high-frequency fields: 10 kHz - 300 GHz. European prestandard prENV 50166-2. Brussels: CENELEC; Jan 1995.

[7] ARIB. Radiofrequency exposure protection standard. ARIB; 1993.

[8] Kuster N, Balzano Q, & Lin JC, editors. Mobile communications safety. London: Chapman & Hall; 1997.

[9] Schmid T, Egger O, & Kuster N. Automated E-field scanning system for dosimetric assessments. IEEE Transactions on Microwave Theory and Techniques 1996 Jan; 44(1): 105-113.

[10] Kuster N, Kastle R, & Schmid T. Dosimetric evaluation of handheld mobile communications equipment with known precision. IEEE Transactions on Communications 1997 May; 80(5): 645-652.

[11] Pokovic K, Schmid T, Christ A, & Kuster N. Design and characterization of E-field probes for lossy media. 1998 (submitted).

[12] NIS81 NAMAS. The treatment of uncertainty in EMC measurement. Teddington, Middlesex: NAMAS Executive, National Physical Laboratory; 1994.

[13] Taylor BN & Kuyatt CE. Guidelines for evaluating and expressing the uncertainty of NIST measurement results. Gaithersburg, MD: NIST; 1994.

2

DOSIMETRY STUDIES ON IN VITRO PREPARATIONS EXPOSED TO 837 AND 1900 MHZ IN A TEM CELL

A.W. Guy

Introduction

This paper describes a new theoretical analysis of the transverse electromagnetic (TEM) cell exposure system used by Wireless Technology Research, LLC (WTR) for in vitro studies of the effect of cellular telephone (CT) and Personal Communication System (PCS) electromagnetic (EM) fields on living cells. This type of TEM cell, previously developed at the University of Washington prior to the WTR program, was chosen since the cross- sectional aspect ratio between the width and height provided a more uniform field and efficient energy coupling to cell flasks than that of the more conventional devices. Also incorporated in the design was a very stable environmental control system suitable for maintaining a very constant temperature, essential for in vitro exposure studies. This model and an earlier analysis of the associated dosimetry was previously described by Guy et al. (1999). Commercially available XFDTD version 4.04 finite-difference time-domain (FDTD) software was used to carry out the earlier analysis. The earlier work was based on modeling the TEM cell without the tapered transition sections, required in actual use for matching the coaxial cable at the feed end and the termination at the load end of the device. This option used less FDTD space by modeling only the inside of the constant rectangular cross section of the TEM cell. The FDTD space included only the volume corresponding to the internal full size cross-sectional area multiplied by the length (20 cm used for the study) of the center section of the cell where the exposed object is placed. The inside of the four walls of the cell were simulated by perfect electric conducting (PEC) boundary conditions, and perfectly matched layer (PML) boundaries were assumed at each end of the TEM cell section. A Laplacean-derived quasi-static electric field (E-field) distribution based on the input voltage was assumed at a cross-sectional plane located at a distance of 15 cells from one end of the TEM cell section.

The new work reported in this paper was dictated by two major considerations. The first was the fact that the magnitude of the calculated electric field distribution over the excitation plane used in earlier models of the empty TEM cell was not consistent with the expected value based on the applied voltage. Though the vendor of the code incorporated a correction factor to account for this in the empty TEM cell in later versions of the code, there was no guarantee that the correction was valid for a loaded TEM cell. Some slight inconsistencies in results of subsequent work with different versions of the code appears to bear this out. For example, calculations using the truncated model with new versions of the code reported in this paper differ slightly from the results reported previously by Guy et al. (1999). The second consideration for the new studies is that they must also address higher frequency PCS exposures where the truncated TEM cell model is no longer appropriate. At CT frequencics the TEM cell will support only the dominant TEM mode. At PCS frequencies, however, higher order TM and TE modes, in addition to the TEM mode, can exist since they are no longer cut off in the full cross section of the TEM cell. However in the narrower cross section at some location in the tapered sections, higher order modes can no longer exist. The waves associated with these modes will reflect back and forth between the opposite tapered sections of the transmission line, producing a complex field distribution unique to the dimensions of the line. Thus a FDTD model of the full TEM cell was developed to calculate these fields and to eliminate the problems associated with the electric field excitation plane in the truncated model.

Full Size TEM Cell Model

The mesh size of the FDTD space containing the new full sizc TEM cell model shown in Figure 1 is 1 mm, with dimensions of 157x101x587 voxels in the x, y, and z directions corresponding to a total volume of 9.3 million cells. This volume is three times greater than that used for meshing the truncated TEM cell. The x, y, and z axes corresponded to the directions across the widest dimension from left to right, the narrowest dimension from left to right, and the longest dimension from bottom to top, respectively, in Figure 1. A feed voltage of 1 V in series with 50 Ω was inserted in the one-cell-wide air gap between the bottom end of the septum and the closed end of the tapered outside shell of the TEM cell. A 50-Ω termination resistance was inserted in the gap at the upper end of the TEM cell.

Figure 2 illustrates the 0.333-mm local meshed FDTD model (external local mesh not shown) of the Falcon #2050 14-ml flask chosen for the in vitro studies. For the studies the flasks contained 10 ml of solution of one of three possible cell groups: suspended blood cells, lymphoma cells, or salmonella cells. For brevity only the blood cells are discussed in this paper. Figure 2 clearly shows the individual voxels and the meniscus at the surface of the solution. Though the meniscus can have considerable influence on the specific absorption rate (SAR) in exposed cells within and near it, for this case its influence was probably negligible since all of the cells were centrifuged down to the lowest 1/3-ml volume of the flask before they were exposed.

For the CT exposures, the vessels were exposed in two groups of three tubes,

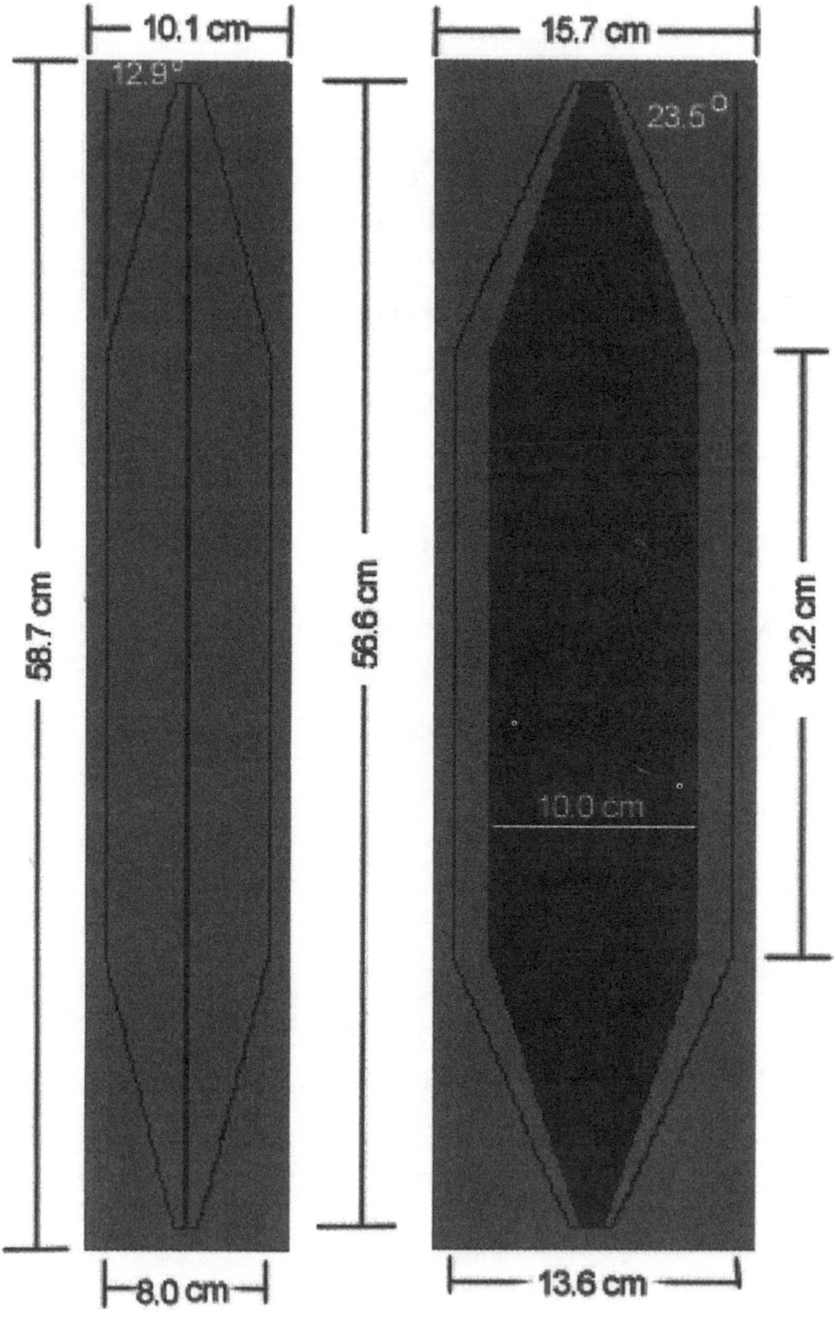

Figure 1. XFDTD model of complete TEM cell used for in vitro studies.

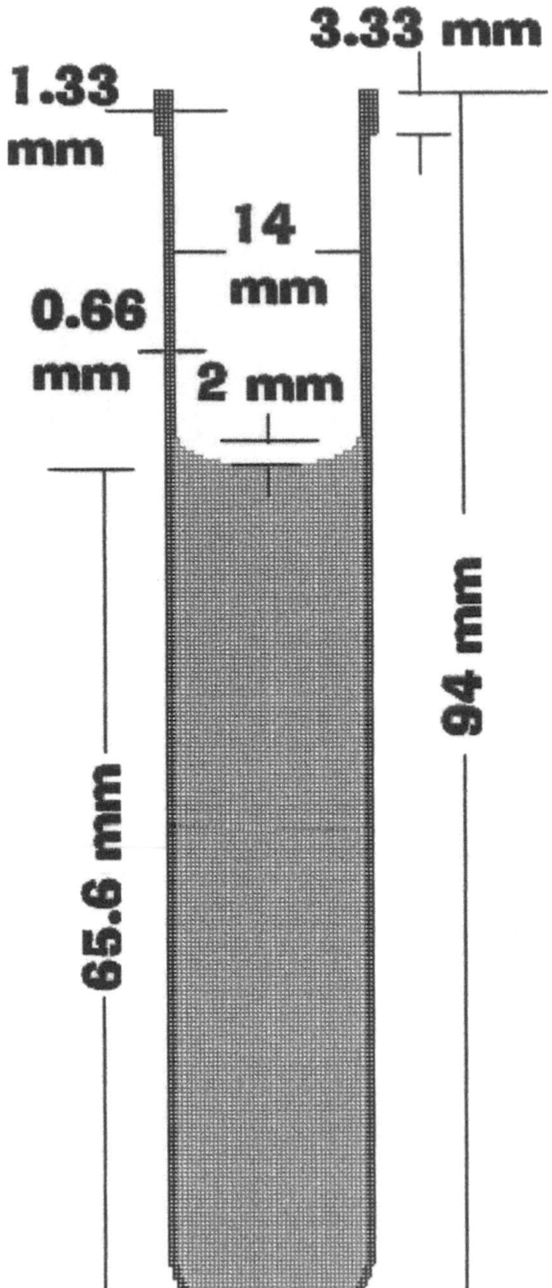

Figure 2. XFDTD model of Falcon #2050 14-ml flask used for in vitro studies.

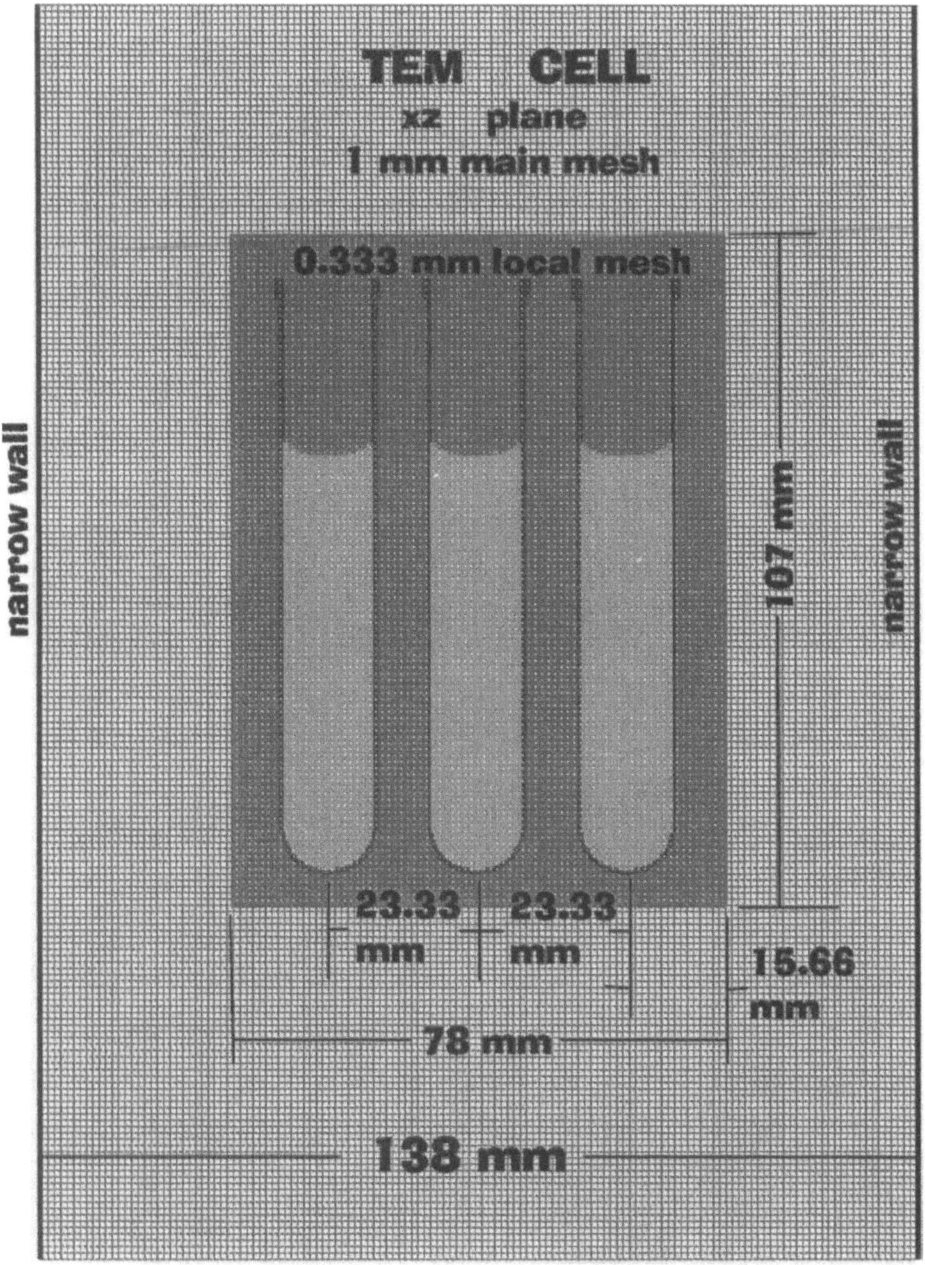

Figure 3. Position of 234x72x321 0.333-mm cell local meshes (each side of septum) in plane parallel to septum.

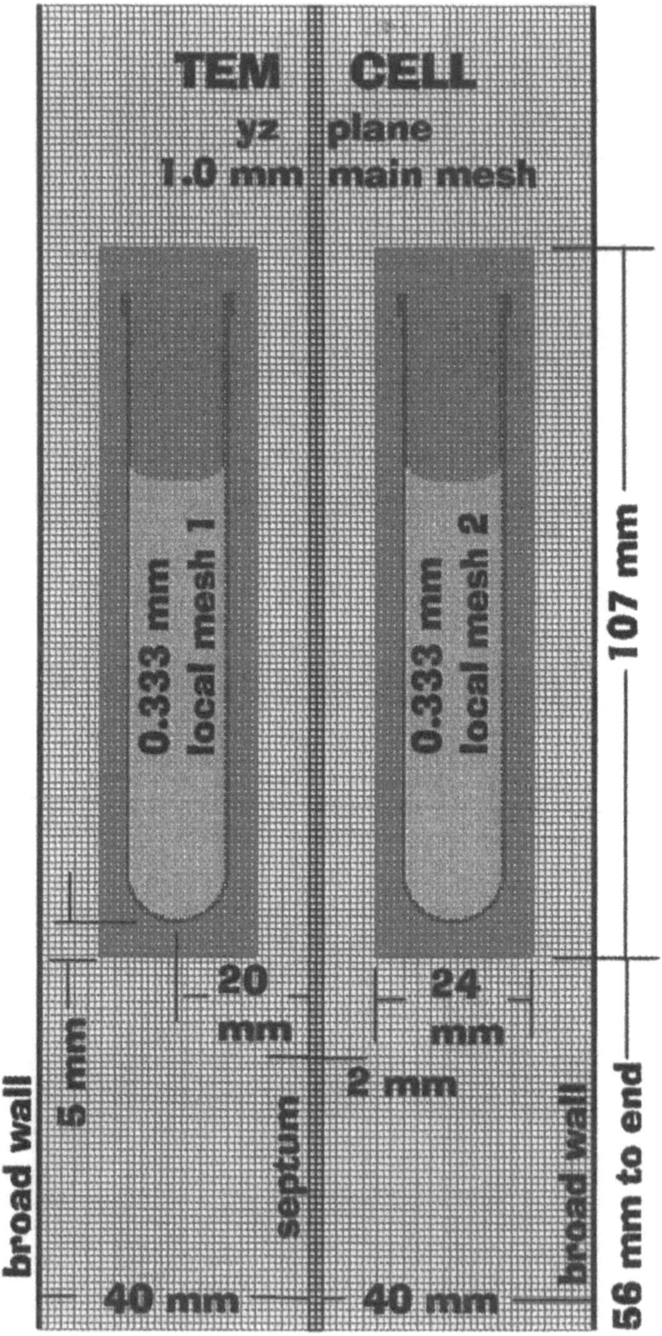

Figure 4. Position of 234x72x321 0.333-mm cell local meshes (each side of septum) in plane perpendicular to septum.

with a group placed on each side of the septum of the TEM cell, as shown in Figure 3 for the xz plane view and in Figure 4 for the yz plane view. These figures illustrate the configuration of the flask group in the 234x72x321 1/3-mm local (dark) mesh contained within the 157x101x587 1-mm main (light) mesh.

Calculation of Fields in Empty TEM Cell

Before calculating the SAR in the flasks exposed at CT and PCS frequencies, FDTD calculations were done to determine the electric and magnetic field uniformity within the TEM cell. Figure 5 illustrates gray scale graphs of the calculated E-field in various planes at major cross sections through the center of the TEM cell. As can be seen in the figure, the E-fields shown in the three principal planes were found to be very uniform in the regions where the flasks would be placed. A maximum E-field strength of 270 V/m was calculated in the exposure volume for a 837 MHz input power of 1 W to the TEM cell. Each increase in darkness of the shade of gray in the figure, starting at the reference of 0 dB for 270 V/m, corresponds to a 1 dB decrease in E-field strength as shown by the legend. Figure 6 illustrates the corresponding magnetic field (H-field) plots for 837 MHz input power of 1 W. The maximum H-field strength in the exposure volume for this case is 0.716 A/m, which corresponds to 0 dB on the legend in the figure.

Figures 7 and 8 illustrate the gray scale graphs of the calculated E-field and H-field at the major TEM cell cross sections for a PCS frequency of 1.9 GHz at 1 W input power. One may note from the graphs for this case there is considerable inhomogeneity in the field distributions. The maximum E- and H-field strengths corresponding to 0 dB for this case were 440 V/m and 0.987 A/m, respectively. Note that the increments between shades of gray for this case is 3 dB, three times greater than that of increments used for the 837 MHz graphs.

SAR Distributions in Exposed Cell Suspensions

The description of the physical characteristics of the TEM cell exposure system with the long axis of the TEM cell oriented in the vertical position has been discussed (Guy et al., 1999). Figure 9 illustrates the gray scale graphical results of the SAR calculations for one of the two identical arrays of tubes exposed to 1 W at 837 MHz, each containing 10 ml of suspended human blood cells. For this case the full-size TEM cell model was used. It was found that the magnitude of the SAR was approximately 19-23% less than what was found using the truncated TEM cell model discussed later in this paper. The exposure field vectors and the SAR distribution are shown for each principal cross section in the vertical direction and horizontal cross sections near the bottom, middle, and top of the tube. The highest SAR occurring at the periphery of the cross-sectional plane shown in the E-K plane at the right side of the figure is due to eddy currents flowing around the periphery of the tube, induced by the magnetic field vector perpendicular to the plane. The magnitude of the SAR at the center of the tube is more than 12 dB (15.8 times) less than at the sides. The

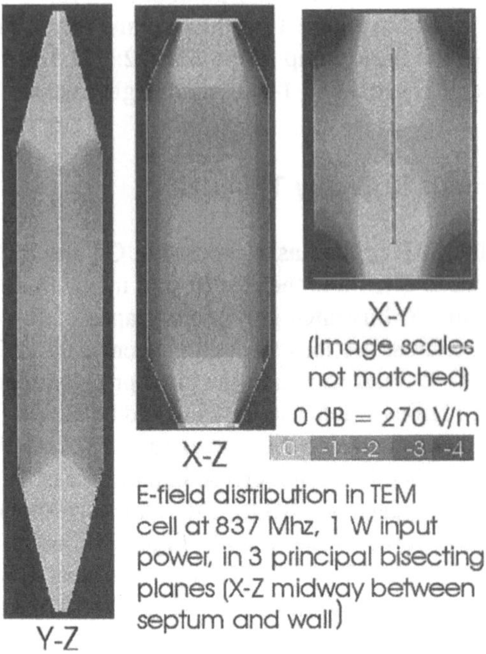

Figure 5. E-field distribution in TEM cell model at 837 MHz calculated by XFDTD.

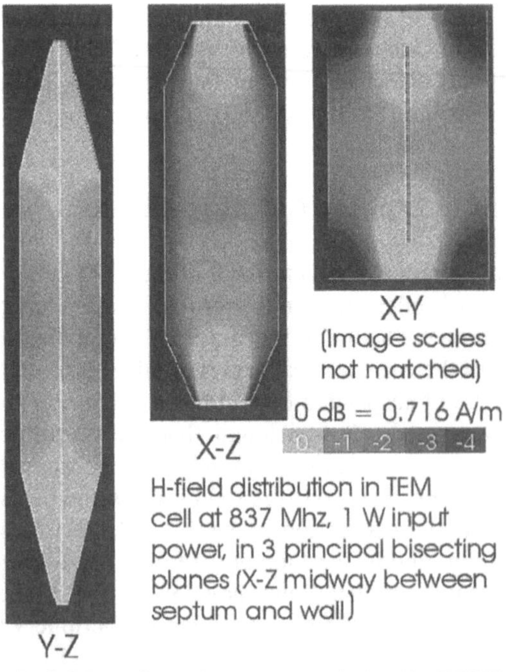

Figure 6. H-field distribution in TEM cell model at 837 MHz calculated by XFDTD.

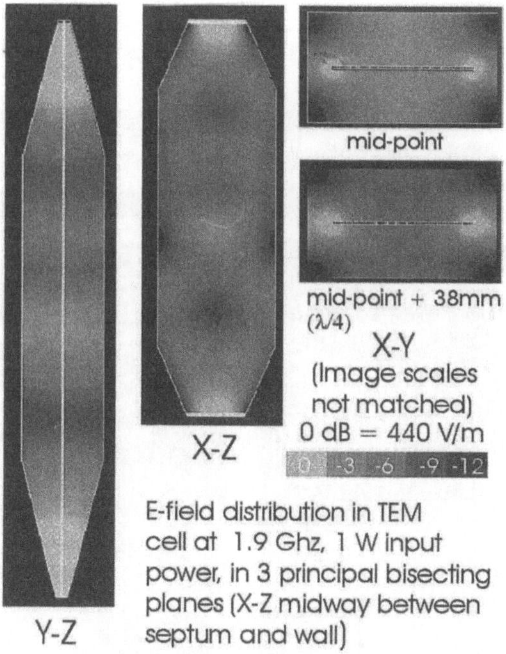

Figure 7. E-field distribution in TEM cell model at 1.9 GHz calculated by XFDTD.

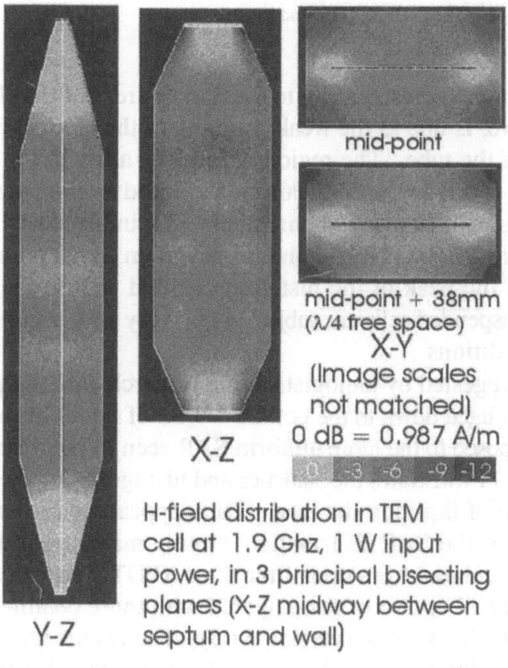

Figure 8. H-field distribution in TEM cell model at 1.9 GHz calculated by XFDTD.

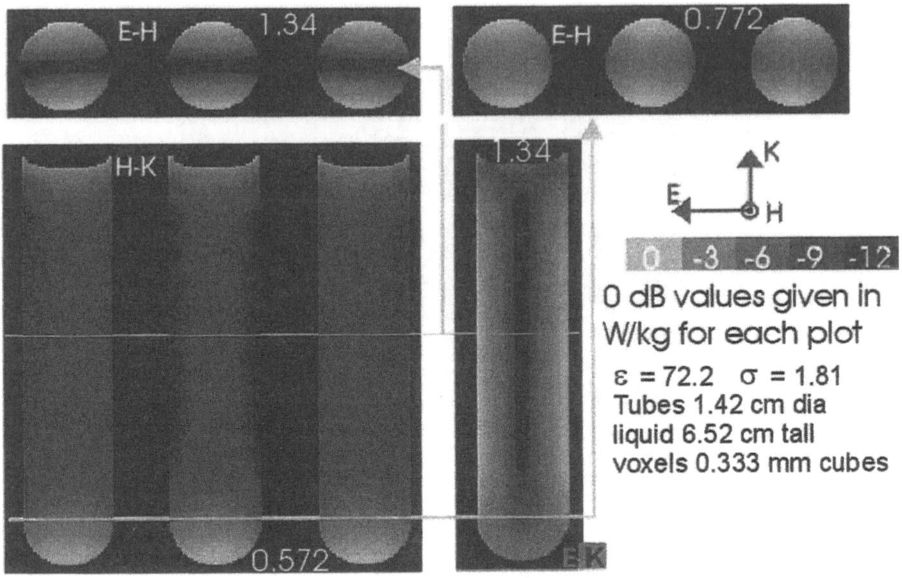

Figure 9. Calculated SAR distribution in full-size TEM cell model in three of a group of six vessels, each containing 10 ml of blood cell medium exposed to 837 MHz in TEM cell with 1 W input power (calculated using full-size TEM cell XFDTD model).

low SAR at the center, corresponding to the darker areas of the H-K cross section at the left of the figure, is due to the weak coupling of the electric field, perpendicular to the long axis of the tube. The region of brighter areas in the H-K section at the bottom of the tubes is due to the eddy currents induced by the magnetic field flowing across the bottom in a horizontal direction. As indicated by the statistics and histogram in Figure 10 (i.e., large standard deviation, large ratio between peak and average SAR, and the peak of the histogram shifted well to the left of the normal distribution), any suspended cells are subjected to a very wide range of SAR values for these exposure conditions.

It has been suggested by biologists doing research with the preparations that if the cells were centrifuged down to the bottom 1/3 ml of liquid in the tube they may be restricted to and exposed to the more uniform SAR seen in the distribution pattern for that region. Figure 11 illustrates the statistics and histogram for the SAR distributions in the bottom 1/3 ml of liquid in the tube. There appears to be marked improvement in the uniformity of the SAR distribution, closely matching a normal curve. The statistical results calculated from one of the later XFDTD codes (version 4.07) using the truncated TEM cell option were compared with those obtained by full size TEM cell option using the latest code (version 5.05). The results are shown in Figure 12 for the full 10 ml solution and in Figure 13 for the lowest 1/3 ml of solution. The SARs from the latter are 19-23% lower than those from the former.

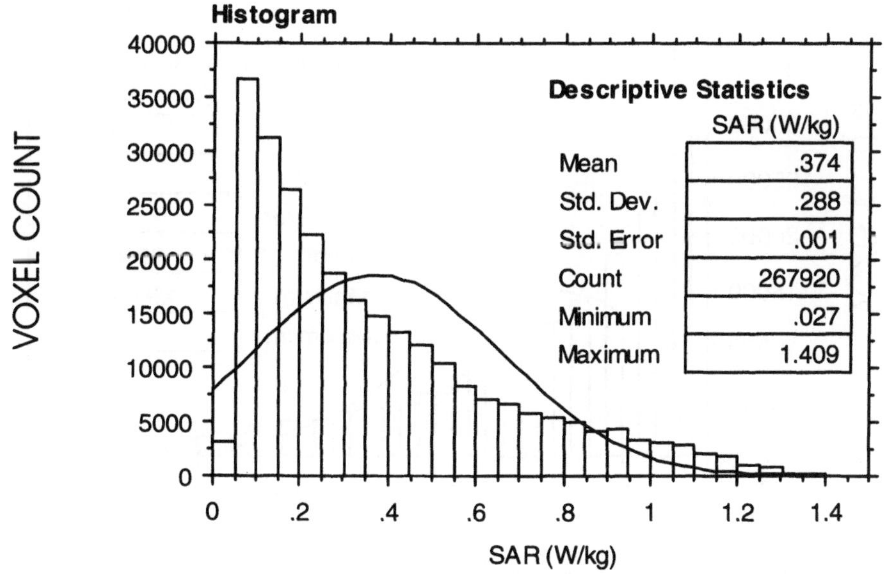

Figure 10. Histograms of calculated SAR distribution in full-size TEM cell model of suspended human blood cells in 10 ml solution with normal comparison in entire volume of first of a group of three tubes exposed to 837 MHz, 1 W input.

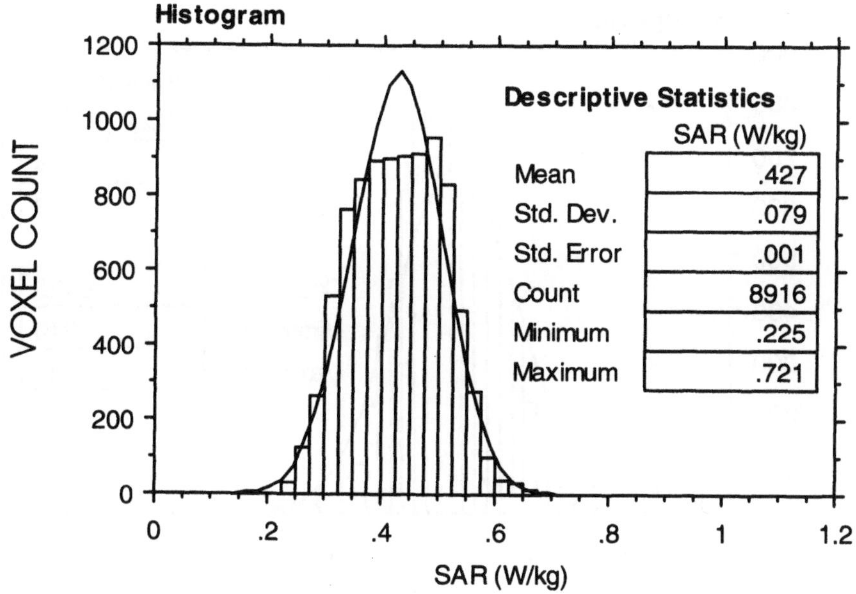

Figure 11. Histograms of calculated SAR distribution in full-size TEM model of suspended human blood cells in 10 ml solution with normal comparison in bottom 1/3 mm of first of the group of three tubes exposed in TEM cell at 837 MHz, 1 W input.

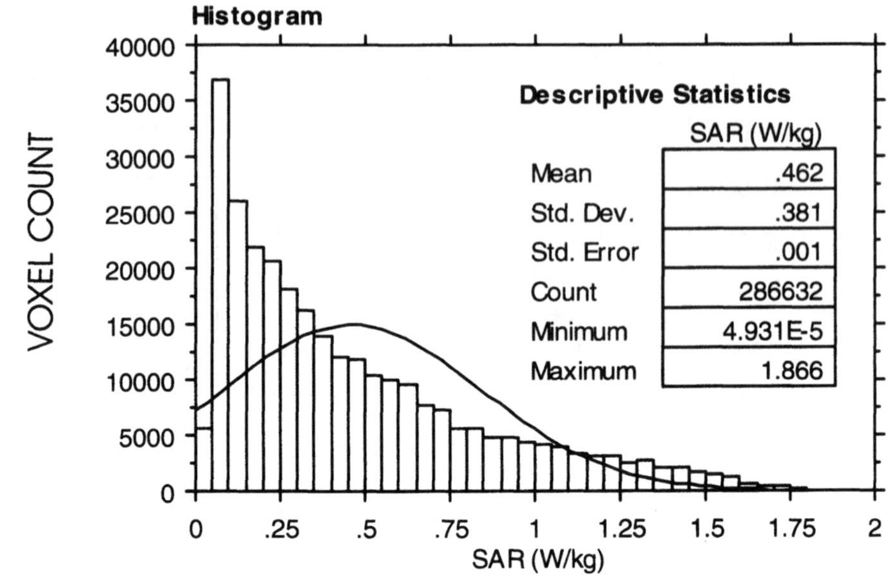

Figure 12. Histograms of calculated SAR distribution in TEM excitation plane model of suspended human blood cells in 10 ml solution with normal comparison in entire volume of first of a group of three tubes exposed in TEM cell at 837 MHz, 1 W input.

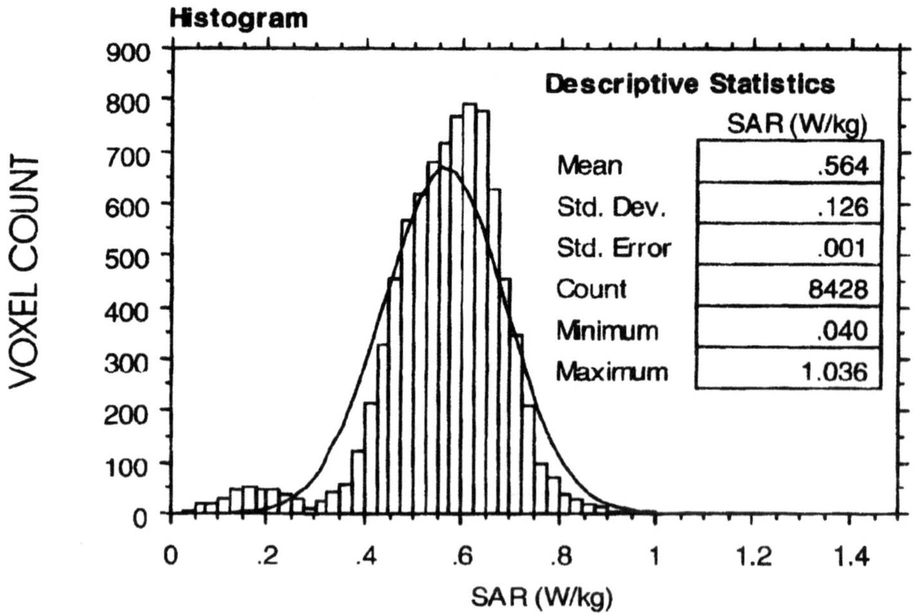

Figure 13. Histograms of calculated SAR distribution in TEM excitation plane model of suspended human blood cells in 10 ml solution with normal comparison in bottom 1/3 mm of first of a group of three tubes exposed in TEM cell at 837 MHz, 1 W input.

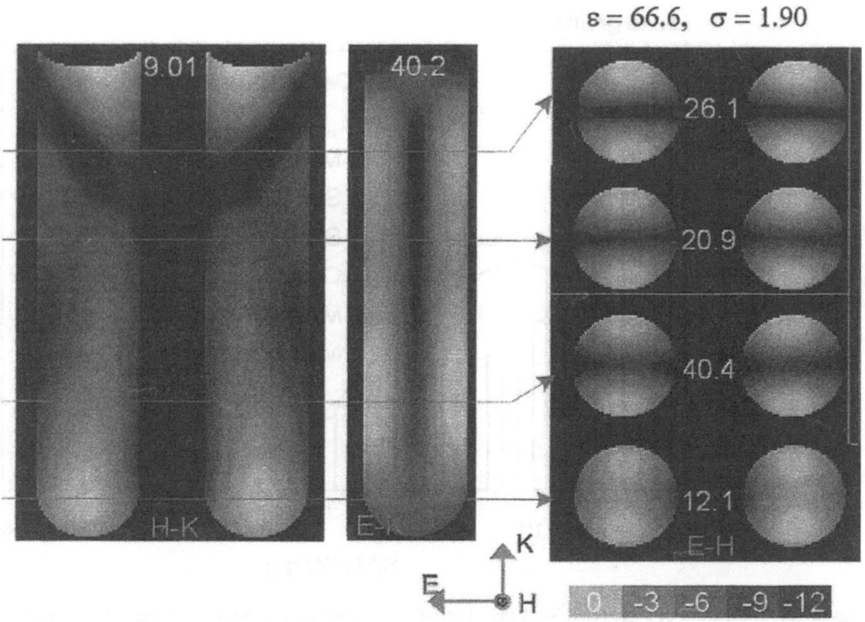

Figure 14. XFDTD-derived SAR distribution in full-size TEM model. Two of a group of four Falcon #2056 14 ml test tubes, each containing human blood cells in 10 ml of medium exposed 1.9 GHz in a TEM cell with 1 W input power (using complete TEM cell model, light lines show positional relationship between cross sections).

Figure 14 illustrates the gray scale graphical results of the SAR calculations for one of the two identical arrays of tubes exposed to the PCS frequency of 1.9 GHz, each containing 10 ml of suspended human blood cells. Again for this case the full-size TEM cell model was used. It can be seen that the calculated SAR distribution is in some ways similar but in other ways different from that obtained for the 837 MHz exposures. The circulating eddy current phenomenon is still apparent from the E-K and E-H plane views of gray scale plots. The SAR distribution in the H-K plane contains bands of low SAR due to the interference patterns set up by the higher order modes in the TEM cell. It can be seen, as expected, that the magnitude of the SAR has increased nearly an order of magnitude due to the increased field strength from constructive multimode interference and the increased coupling at the higher frequency. As for the case of the 837 MHz exposures, the statistics and histogram in Figure 15 (i.e., large standard deviation, large ratio between peak and average SAR, and the peak of the histogram shifted well to the left of the normal distribution) show any suspended cells would be subjected to a very wide range of SAR values for these exposure conditions. However for the cells centrifuged down to the lower 1/3 ml of solution, the 1.9 GHz exposure results in about the same uniformity as obtained with the 837 MHz exposure as can be seen from the histogram in Figure 16.

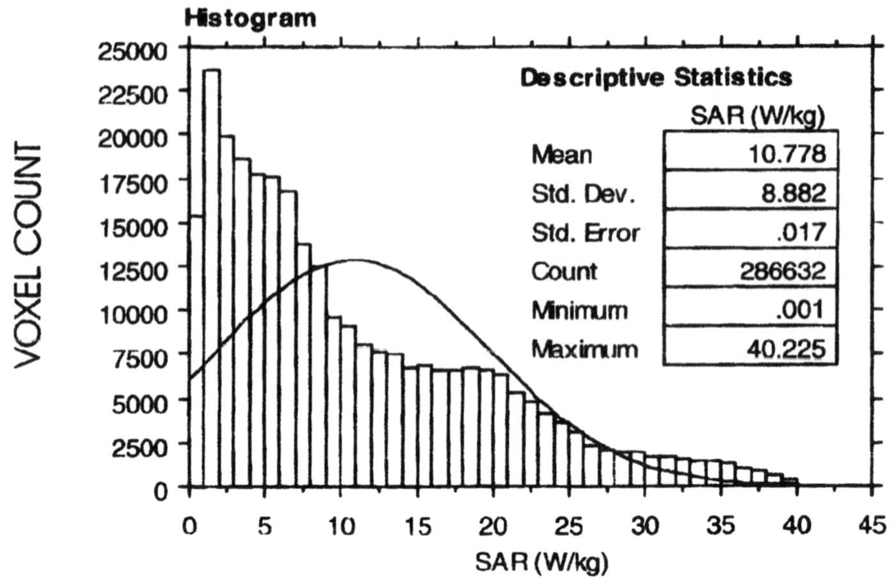

Figure 15. Histogram of calculated SAR distribution in full-size TEM model of suspended human blood cells in 10 ml solution with normal comparison in entire volume of one of a group of four tubes exposed in TEM cell at 1.9 GHz, 1 W input.

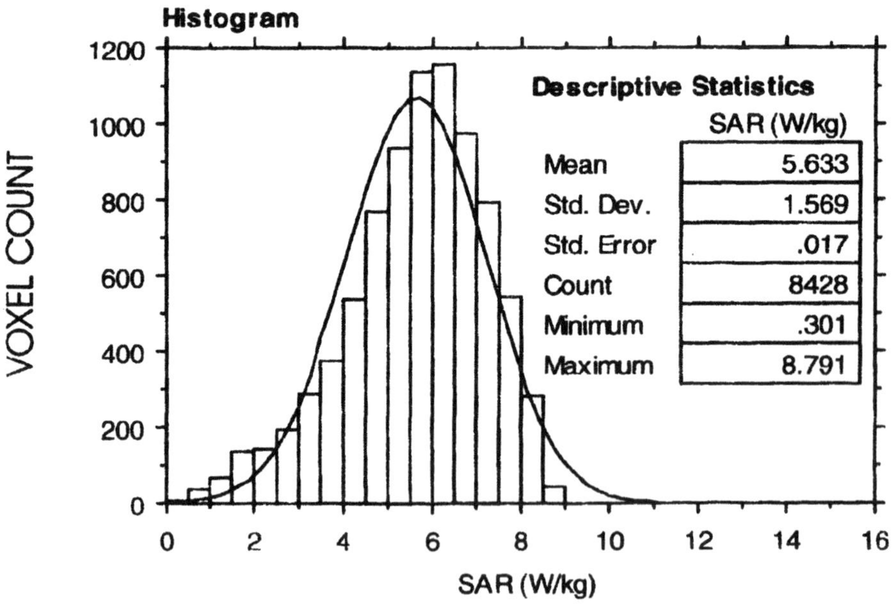

Figure 16. Histogram of calculated SAR distribution in full-size TEM model of suspended human blood cells in 10 ml solution with normal comparison in first 1/3 ml from bottom of one of a group of four tubes exposed in TEM cell at 1.9 GHz, 1 W input.

Conclusions

A full-size FDTD TEM cell model has been formulated and used to calculate the SAR distribution patterns in cell suspensions exposed to 837 MHz and 1.9 GHz electromagnetic fields. Regardless of whether exposures are from a single dominant TEM mode or multiple modes, the results provide quantitative dosimetry support for in vitro cellular studies relating to the assessment of the safety of cellular telephones and PCS in terms of human health effects. The results provide guidance as to the best location for the placement of the cells in the exposed solution and for the setting of input power levels to the TEM cell exposure system. The results indicate that the calculated SARs based on the voltages and load options using the full-size TEM cell model are approximately 20% lower than values obtained by the truncated TEM cell option of the XFDTD code. It is assumed that the calculations obtained from the realistic full-size model provide more accurate results than those obtainable from the truncated model.

Reference

Guy AW, Chou CK, & McDougall JA. A quarter century of in vitro research: A new look at exposure methods. Bioelectromagnetics 1999; 20 (suppl 4): 21-39.

3

A RAT-HEAD EXPOSURE SYSTEM FOR EXPOSURE TO RF FIELDS FROM HANDHELD WIRELESS TELEPHONES

D.I. McRee, C.K. Chou, J.A. McDougall, K.W. Chan, & A.W. Guy

Abstract

A rat-head exposure system has been developed for Wireless Technology Research, LLC (WTR) through a contract with Dr. C.K. Chou who was at the City of Hope National Medical Center during the development of the system. In the first phase of the development of the system the finite-difference time-domain (FDTD) method was used to calculate the specific absorption rates (SARs) in ellipsoidal rat models with simulated fields from different loop antennas located at various distances from the models. From this analytical work a 3 cm x 1 cm rectangular loop was designed with optimized energy coupling and impedance matching.

Sprague Dawley rats were then exposed using the 3 cm x 1 cm loop antennas, tuned either to 837 MHz or 1957 MHz, for the determination of SAR distribution utilizing the thermographic technique. In addition, point SARs in the brains of restrained rats were measured using fiber optic probes. The calculated and measured SAR patterns were in general agreement. The average brain-to-whole body SAR ratio was 20 to 1 for both frequencies. At 837 MHz, the maximum measured SAR in the restrained rat brain was 51 W/kg/W in the cerebellum and 40 W/kg/W at the top of the cerebellum. For a maximum SAR of 10 W/kg, only 0.2 W input power to the loop is required. The average brain SAR in a 300-g medium size rat for 0.2 W input power is estimated to be 4.8 W/kg.

The exposure system has been used to expose Sprague Dawley rats to 837 MHz using analog, TDMA, and CDMA cellular phones as generators of the signals to the loop antennas. At this time, the longest duration of exposure has been 3 hours.

Introduction

At the beginning of the Wireless Technology Research, LLC (WTR) program, it was necessary to develop a system for the exposure of animals which would simulate as closely as possible the exposure of humans to cellular telephones. On studying the requirements of such systems, it was decided by experienced people in the field that the system should have the following characteristics:

1. The system should provide energy deposition in the animal brain that will

allow the study of the effects of a wide range of specific absorption rates (SARs) with minimal whole-body SAR;

2. An antenna with high energy coupling and impedance matching is required to obtain SARs without high power input to the antenna;
3. A method for restraining the animal without overly stressing them in order to obtain constant exposure with time is required; and
4. The dosimetry to characterize the SAR distribution in the brains of animals must be determined.

A committee of experts in the field of microwave exposure and dosimetry met three different times to discuss alternative systems and approaches for exposing animals to fields from cellular phones. From the information generated at these meetings it was decided that a system which exposed only the head should be developed since one was not available. A contract was granted to the City of Hope National Medical Center, with Dr. C.K. Chou as Project Director, for the purpose of developing the animal exposure system. The details of the completed system have been previously published (Chou et al., 1999). A brief description of the steps to design, assemble, and calibrate the system will be presented in this paper.

System Development

The first step in the development of the system was to design an antenna that would deliver high SARs with low power input to the antenna. The finite-difference time-domain (FDTD) analytical method (Kunz & Leubbers 1993; Taflove 1995) was used to model dipole and loop antennas to see which was best for localized rat-head exposure. Using a muscle ellipsoid model about the size of a rat, FDTD calculations were made to choose the best antenna design. The results of these calculations are

Loops (cm)	Spacing (mm)	Impedance (Ω)	Relative SAR
3x1	5	5.6 + j 426	1.0
2x2	5	5.7 + j 473	0.67
2x3	5	11.9 + j 746	0.50
2x2	2	8.13 + j 474	2.56
2x2	11	3.86 + j 472	0.12
2x3	2	16.6 + j 749	1.82
3x1	2	8.63 + j 429	3.84
3x0.5	2	3.87 + j 294	4.28
3x 0.2	2	1.51 + j 185	3.82

(Muscle Ellipsoid: ϵ=51, σ=1.6 S/m, ρ=965 kg/m^3; frequency = 915 MHz)

Table 1. FDTD calculated impedance and relative SAR changes as a function of loop size and spacing.

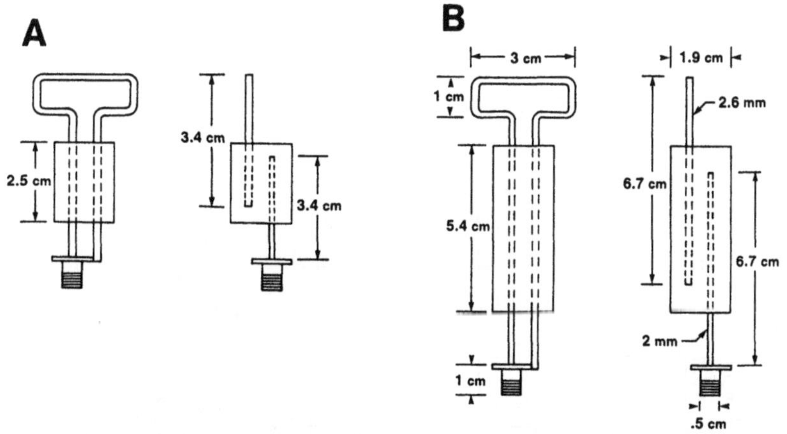

Figure 1. Loop antenna for exposing rat head (A) at cellular phone frequencies (800-900 MHz) and (B) at PCS frequencies (1800-2000 MHz). (Reprinted from McRee et al. Development of a rat head exposure system for simulating human exposure to RF fields for handheld wireless telephones. Bioelectromagnetics 1999, Wiley-Liss, Inc., a subsidiary of John Wiley & Sons, Inc.)

shown in Table 1. From these calculations a loop antenna with dimensions of 3 cm x 1 cm spaced 5 mm from the ellipsoid was chosen for further testing. The design of the loop for 800-900 MHz is shown in Figure 1A and the loop for 1800-2000 MHz is shown in Figure 1B.

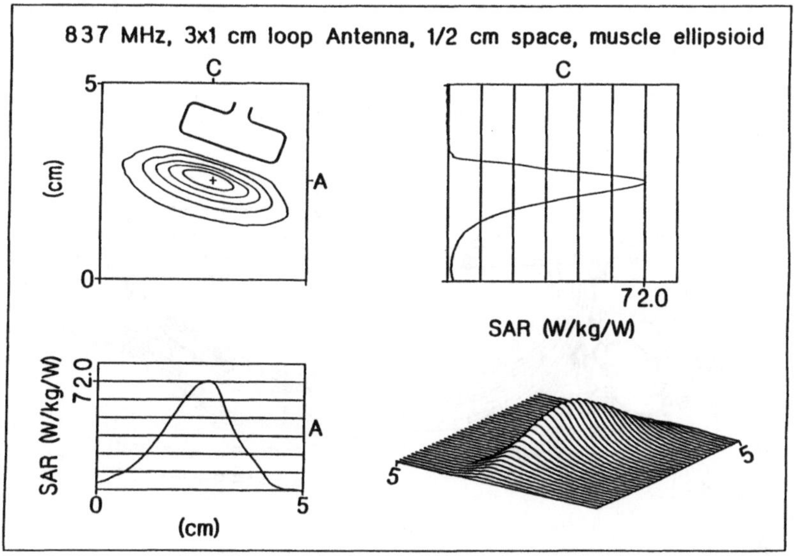

Figure 2. Thermograms of a muscle ellipsoid (20 cm long and 5.6 cm in diameter) exposed to 837 MHz fields with the 3 cm x 1 cm loop. The upper left is a contour plot showing six levels of temperature elevation. The "+" indicates the peak SAR location. Line scans (A and C) through the peak are in lower left and upper right, respectively. A profile scan is shown at the lower right. Normalized SAR data can be obtained from the line scans. (Reprinted from McRee et al. Development of a rat head exposure system for simulating human exposure to RF fields for handheld wireless telephones. Bioelectromagnetics 1999, Wiley-Liss, Inc., a subsidiary of John Wiley & Sons, Inc.)

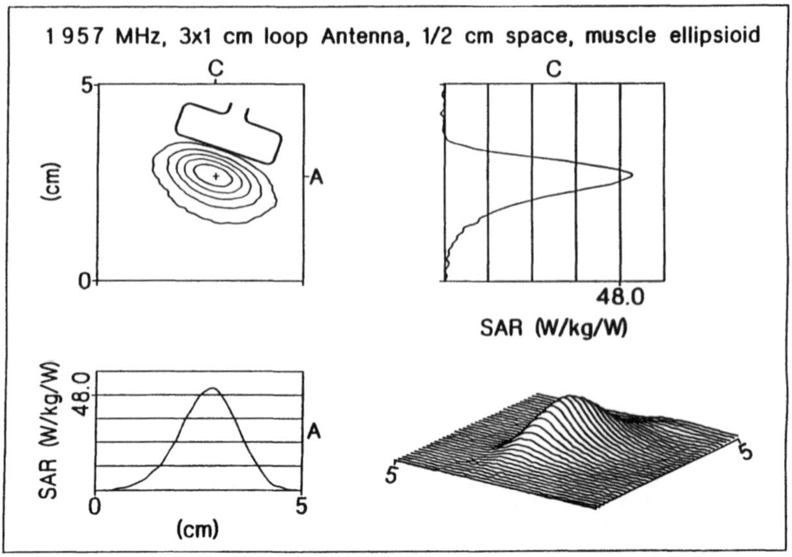

Figure 3. Thermograms of a muscle ellipsoid (20 cm long and 5.6 cm in diameter) exposed to 1957 MHz fields with the 3 cm x 1 cm loop. The upper left is a contour plot showing six levels of temperature elevation. The "+" indicates the peak SAR location. Line scans (A and C) through the peak are in lower left and upper right, respectively. A profile scan is shown at the lower right. Normalized SAR data can be obtained from the line scans. (Reprinted from McRee et al. Development of a rat head exposure system for simulating human exposure to RF fields for handheld wireless telephones. Bioelectromagnetics 1999, Wiley-Liss, Inc., a subsidiary of John Wiley & Sons, Inc.)

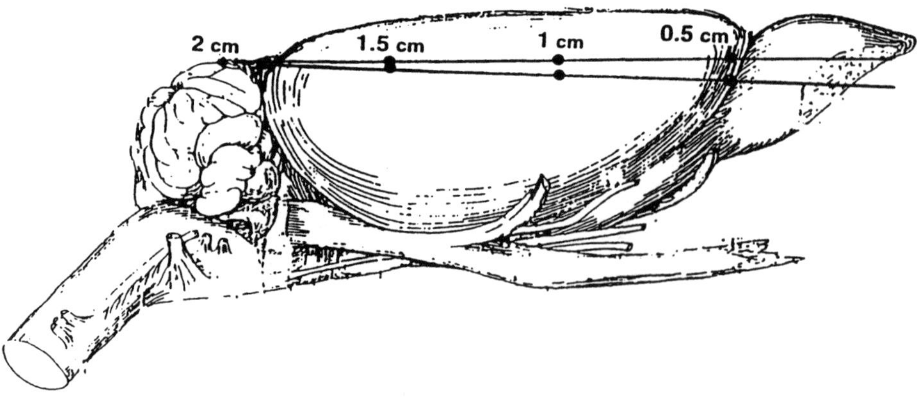

Figure 4. Catheter tracks during point SAR measurement. Rats #7 & #10 at upper track and Rats #8 & #9 at lower track. (Reprinted from McRee et al. Development of a rat head exposure system for simulating human exposure to RF fields for handheld wireless telephones. Bioelectromagnetics 1999, Wiley-Liss, Inc., a subsidiary of John Wiley & Sons, Inc.)

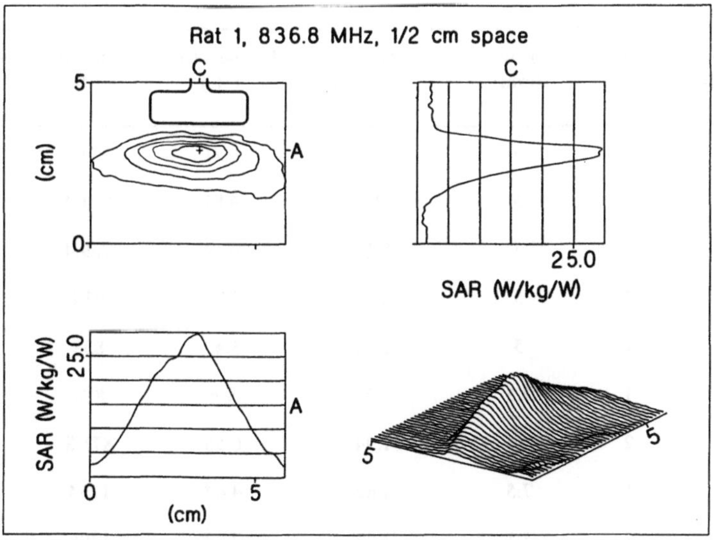

Figure 5. Thermogram of Rat #1 exposed to 837 MHz fields using the 3 cm x 1 cm loop antenna. The upper left is a contour plot showing six levels of temperature elevation. The "+" indicates peak SAR location. Line scans (A and C) through the peak are in lower left and upper right, respectively. A profile scan is shown at the lower right. Normalized SAR data can be obtained from the line scans. (Reprinted from McRee et al. Development of a rat head exposure system for simulating human exposure to RF fields for handheld wireless telephones. Bioelectromagnetics 1999, Wiley-Liss, Inc., a subsidiary of John Wiley & Sons, Inc.)

After selection of the antennas using the FDTD method was completed, the antennas were experimentally tested on a muscle ellipsoid model. Thermograms of antennas antennas were experimentally tested on a muscle ellipsoid model. Thermograms of the exposure of the ellipsoid model to the radio frequency (RF) fields from the loop are shown in Figures 2 and 3. These results show that there is excellent coupling of power from the antenna to the ellipsoid model of the rat as was predicted by the FDTD calculations.

After the design of the antenna using the muscle ellipsoid, the next step was to determine the absorption in the rat brain with power input to the antenna. Sprague Dawley rats were used in these studies because the first scheduled experiments with the system were to use this species. Point SARs were determined in the rat brains by inserting a Teflon tube into the brain of dead rats and measuring the temperature at various points along the tube (Figure 4). Temperatures were measured along the Teflon tube at 5-mm intervals starting at the brain anterior. The SARs at these points were calculated from the time-temperature profiles and used in the validation of the absorption distribution.

Thermograms were also taken in the rat brain. The rat was frozen, cut in half, put back together, and thawed. The rat was then exposed to high levels of microwave power and immediately separated so that a thermogram could be taken. The SARs through the center of the brain along the vertical cut are shown in Figure 5 for 836.8 MHz. A summary of the thermographically determined peak SARs in Sprague Dawley rat brains exposed to cellular telephone frequency fields with the loop

Frequency (MHz)	Rat	Distance (mm)	Distance (VSWR)	Temperature (°C)	Energy (J)	Maximum SAR (W/kg/W)
1965	1	5	1.1	8.71	523.4	59.9
842	1	5	2.2	4.81	521.4	33.2
842	2	5	2.15	6.14	759.4	29.1
837	1	5	2.0	7.71	934	29.7
834.5	2	5	2.65	5.49	838	23.6
837	1	2.5	1.74	9.4	886	38.2
837	2	2.5	1.94	6.58	871.8	27.2
837	1	7.5	2.62	4.42	1254	12.7
837	2	7.5	2.52	7.07	1306	19.5

Table 2. Thermographically determined peak SARs in Sprague Dawley rat brains exposed to cellular telephone frequency fields with loop antennas.

antenna is presented in Table 2.

Since temperature probes can only determine SARs at points and the thermographic method along a plane in the brain, the FDTD method was used to provide three-dimensional volumetric information. A histogram of the SARs in each

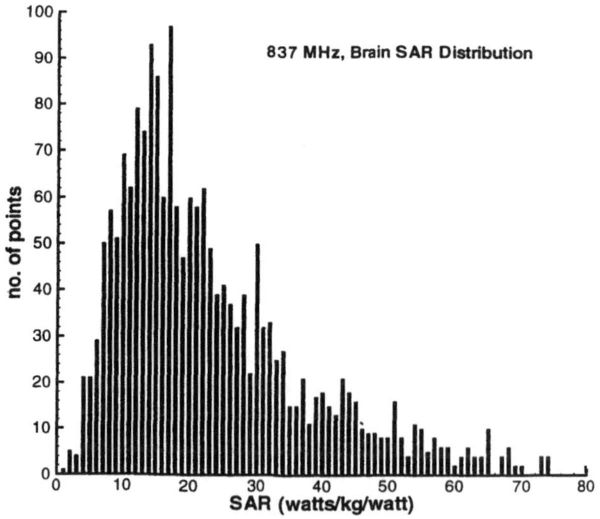

Figure 6. Histogram of the pixels in the brain of the rat model exposed to 837 MHz fields with the loop antenna. (Reprinted from McRee et al. Development of a rat head exposure system for simulating human exposure to RF fields for handheld wireless telephones. Bioelectromagnetics 1999, Wiley-Liss, Inc., a subsidiary of John Wiley & Sons, Inc.)

	Maximum	Average	STD	Volume (mm^3)	% Absorption
837 MHz					
Brain	80.6	23.8	14.4	1.922	9.1
Head	223	6.1	11.5	66,548	42.4
Whole body	223	1.2	4.6	543,946	68.7
1957 MHz					
Brain	71.9	22.6	11.3	1,922	8.7
Head	192	7.8	11.1	65,815	53.1
Whole body	192	1.1	4.6	543,977	59.4

Table 3. FDTD calculated SAR (W/kg/W) based on rat CT images.

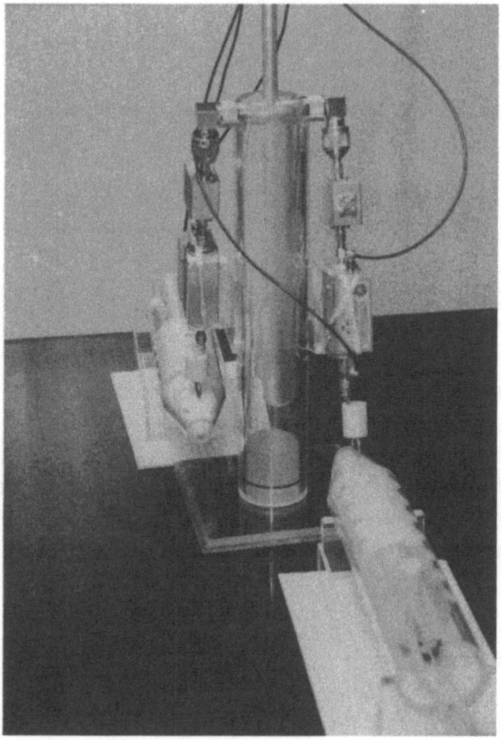

Figure 7. Setup for exposing two rats, including the two-way power splitter, isolators, directional couplers, loop antennas, and rat restrainers. (Reprinted from McRee et al. Development of a rat head exposure system for simulating human exposure to RF fields for handheld wireless telephones. Bioelectromagnetics 1999, Wiley-Liss, Inc., a subsidiary of John Wiley & Sons, Inc.)

Figure 8. Setup for exposing eight rats with computer monitoring system.

pixel inside the brain exposed to 837 MHz is shown in Figure 6. From histograms like this one, the maximum and average SARs in the brain, head, and whole body and the percentage of power absorption can be calculated. These calculated values of SAR are presented in Table 3.

The results of the data show that at 837 MHz the average SAR in a rat brain is about 24 W/kg/W with the 3 cm x 1 cm loop antenna spaced 5 mm from the rat. The ratio of SAR in the brain to whole body is 20 to 1. In the experiments to determine if exposure to cellular phones causes DNA damage to the brain, peak SARs of 10, 5, 1, and 0 W/kg were used. At 837 MHz 0.2 W of input power to the antenna produced a peak SAR of 10 W/kg and an average SAR of 4.8 W/kg in the brain and as low as 0.24 W/kg average whole-body SAR.

Total System Design

This head-only exposure system has been used to study DNA damage in the brain. A photograph of an exposure unit with two rats in position to be exposed is shown in Figure 7. In each leg of the exposure unit is an isolator, a bidirectional coupler, and a loop antenna. The isolator prevents disturbances in the line to affect the power setting to the loop antenna. The bidirectional coupler measures the forward and reflected power, and these power measurements are used to set the desired power to each antenna and to monitor the input power during the total exposure period. The rats are restrained in the conical Plexiglas cones and kept in the cones with a plug at the rear. The antenna is placed against the conical cone above the brain of the animal.

A system for exposing eight animals at the same time is shown in Figure 8. Each stand supports two exposure legs. Variable attenuators in each line provide the capability of setting the input power of each pair of antenna and each pair of antenna can be set at any selected input power. In our experiment the input powers were set so that the rat brains were exposed to peak SARs of 0, 1, 5, and 10 W/kg.

Conclusions

The head-only exposure system has been used for three experiments to study the effects of exposure to analog (FM), Time Division Multiple Access (TDMA), and Code Division Multiple Access (CDMA) phone technologies at 837 MHz on DNA damage in the brains of rats. The animals were conditioned to enter the cone restraint for 3 hours per day for 8 days before being exposed. The animals adapted to the cones in most cases and entered the cones readily after the habituation. The animals were exposed for 3 hours. The peak SARs were 10, 5, 1, and 0 W/kg, and the average SARs over the whole brain were 4.8, 2.4, 0.48, and 0 W/kg. The whole-body SARs were 0.24, 0.12, 0.024 and 0 W/kg. These whole-body SARs are much less than those which thermally stress the animals. This exposure system is the only system currently available that can expose the brains of animals to SARs an order of magnitude higher than SARs in human brains when exposed to cellular telephones without producing thermal stress to the animals. This system provides the capability of conducting dose-response experiments which are important in toxicological testing. The exposure system described above meets all the requirements originally determined to be necessary to conduct experiments in animals that best simulate the exposure to humans. Although the photograph of the system only shows eight animals being exposed, the system can be expanded modularly to expose as many animals as required by experimental design.

References

Chou CK, Chan KW, McDougall JA, & Guy AW. Development of a rat-head exposure system for simulating human exposure to RF fields for handheld wireless telephones. Bioelectromagnetics 1999; 20 (suppl 4): 75-92.

Kunz KS & Leubbers RJ. The finite-difference time-domain method for electromagnetics. Boca Raton, FL: CRC Press, 1993.

Taflove A. Computational electrodynamics: The finite-difference time-domain method. Norwood, MA: Artech House, 1995.

4

A FAST TOUR OF A DECADE OF BIOELECTROMAGNETIC HYGIENE RESEARCH

M. Lundquist

Introduction

Hygiene is the prevention of disease. Bioelectromagnetic hygiene is the profession concerned with preventing those diseases that result from exposure to electromagnetic fields or radiation, especially in the non-ionizing portion of the electromagnetic spectrum. A bioelectromagnetic hygienist is a trained, qualified person who practices bioelectromagnetic hygiene. My own background is that of a Ph.D. physicist and an industrial hygienist who is certified in the comprehensive practice of industrial hygiene by the American Board of Industrial Hygiene.

Industrial hygiene is a profession of disease prevention that became established in the U.S. following the 1911 investigation of Illinois industrial lead-using firms, which Alice Hamilton, M.D., undertook at the request of the Governor of Illinois to document the fact that men were being made ill by the work that they did in the course of performing their jobs. It is a broad-based profession: it prevents those diseases that are caused by exposure to hazardous environmental agents of all kinds, whether these are chemical, physical, or biological agents. By the 1930s, industrial hygienists were working in the US Department of Labor, as well as in many state agencies and in the large industrial firms of that time.

In addition to industrial hygiene, there is another recognized profession of disease prevention, health physics, which was founded in the U.S. in the early 1940s to prevent those diseases caused by exposure to ionizing agents in the environment. One such agent is the ionizing portion of the electromagnetic spectrum. Initially, health physicists confined their attention to the prevention of those diseases resulting from exposure to agents that produced ionization, though in recent years they have tried to broaden the scope of their profession so as to address the health hazards of exposure to non-ionizing electromagnetic radiation as well. However, their efforts have provided no new insights into diseases or mechanisms, perhaps because the specialized knowledge of health physics, being based on ionization, appears to be inapplicable to non-ionizing electromagnetic fields.

Every profession is based on a core of specialized knowledge. What, then, is the scientific foundation for bioelectromagnetic hygiene? It is based on two bodies of physical knowledge, electromagnetic field theory and relativity theory, to which the disease-prevention principles of industrial hygiene have been joined. Electromagnetic field theory is the name given to that body of knowledge developed by both physicists and electrical engineers from the equations of the electromagnetic field attributed to Scottish nineteenth-century physicist James Clerk Maxwell. Relativity theory is a body of knowledge also founded on Maxwell's equations of electromagnetism, but developed exclusively by physicists, beginning with Albert Einstein in the very early twentieth century. The combination of these two bodies of specialized physical knowledge with the principles of industrial hygiene constitutes the specialized knowledge that serves as the scientific foundation underlying the profession of bioelectromagnetic hygiene.

There is nothing in bioelectromagnetic hygiene that restricts its application to any particular portion of the electromagnetic spectrum. However, as an existing profession of disease prevention, health physics, has been effectively addressing the health hazards of *ionizing* electromagnetic radiation for the past half-century, the current need is for a profession dedicated to the prevention of those diseases caused by exposure to *non-ionizing* electromagnetic radiation and fields. For this reason, the non-ionizing portion of the electromagnetic spectrum is the current focus of bioelectromagnetic hygiene.

Disease Prevention via an Exposure Standard: What Standard Protects Health?

One way to prevent disease is to establish appropriate standards governing exposure to non-ionizing radiation. Figure 1 shows three widely different exposure limits that have been proposed for human safety at different times by different parties—governments, non-profit organizations, or individuals—around the world. The values displayed come from Solon.[1] (Others have given somewhat different values for the Soviet standard.)

ANSI is an acronym for the American National Standards Institute, an organization that provides institutional support for the establishment and periodic revision of a variety of voluntary consensus standards. Notice that ANSI C95, the original voluntary consensus health standard for human exposure to non-ionizing radiation in the U.S.—and the one most often relied upon by US corporations and government agencies, since no agency of the US government has established an enforceable standard for such exposure that applies to the private sector—is sponsored by the Institute of Electrical and Electronics Engineers (IEEE), a professional society of technically trained people who are not disease prevention professionals.

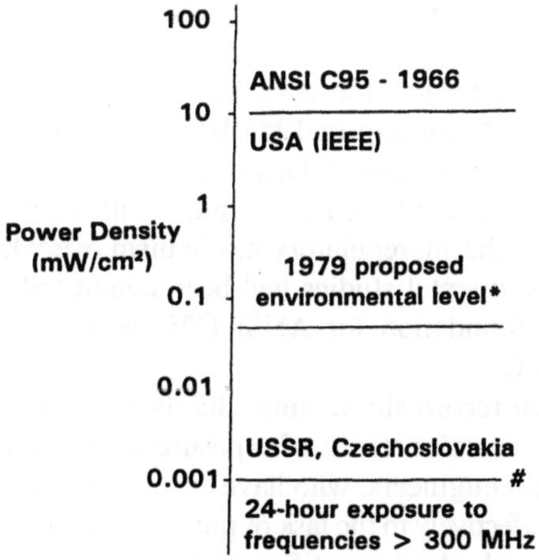

*This environmental level, proposed for New York City and developed by a disease prevention professional, was never actually adopted.
This level was established in 1958 or not long thereafter as a human exposure standard by government officials charged with defining the conditions needed to protect human health.
Figure 1. Historical microwave radiation exposure standards, actual and proposed.[1] This diagram shows that the standards have ranged over four orders of magnitude, indicating the enormous uncertainty that has existed in the past (and that still exists today) regarding a safe level of exposure to microwave radiation.

It is an accident of US history that in the early 1950s, when the health hazards of radar first began to be appreciated and the US military desired to establish health protection standards for exposure to it in order to protect US servicemen, electrical engineers were identified by those in the military as the professionals best suited to meet the challenge of establishing a human exposure standard to protect against microwave radiation.[2]

Electrical engineers were selected because they were the ones who had designed and installed the radar systems on ships and in airplanes and because they knew, and knew how to apply, a good deal of electromagnetic field theory. Certainly the electrical engineering profession does possess certain specialized knowledge that is vitally needed in the task of establishing a human exposure standard to protect against microwave and radio-frequency fields—for example, knowledge of the way electromagnetic fields behave near antennas of various types and an understanding of how to design and use the measuring instruments commonly employed to characterize electromagnetic fields. But this profession should play only a supporting role in the task of developing such exposure standards, letting the final responsibility rest with qualified disease prevention professionals, rather than assuming the responsibility for developing and maintaining a standard that lies so far outside the realm of the professional expertise of an electrical engineer as the protection of human health does.

In hindsight, it is evident that a background in physics, rather than in electrical engineering, would have been preferable for those who originally undertook the task of developing a standard to protect the health of people exposed to microwave radiation. Indeed, the history of the development of the ANSI standard shows that a German-born, German-educated biophysicist, Herman P. Schwan, was probably the single individual most responsible for the development of the original 1966 ANSI standard, ANSI C95.

Dr. Schwan addressed the issue of the thermal hazard associated with microwave radiation, determining in 1953 from theoretical considerations that 10 mW/cm^2 was the value of the power density that, in a worst case scenario, was the maximum allowable in order to ensure that mammals in the far field of a radiation source did not find their thermoregulatory mechanism overwhelmed. Thirteen years later, after many experimental studies had been conducted, his estimate remained valid. He laid the foundation for ANSI C95, which has since been modified considerably by the IEEE.

The historical record shows quite clearly that when the issue is protecting human health from the harmful effects of exposure to electromagnetic radiation, it is physicists, *not* electrical engineers, who have demonstrated the ability to apply their knowledge of physics effectively to the task of protecting human health. The existence of the profession of health physics testifies to this.

Had electrical engineering given rise to a profession dedicated to the prevention of those diseases that result from exposure to non-ionizing electromagnetic radiation, the situation in the world today would be quite different from what it actually is.

Pertinent Equations from Electromagnetic Field Theory

Maxwell's equations of electromagnetism, a mathematical description successfully summarizing and synthesizing the findings of the experimental investigations into electricity and magnetism of the nineteenth century, constitute the foundation of electromagnetic field theory. In their simplest form they are four equations relating the four basic vectors of the electromagnetic field—two electric vectors: **E** (the electric field) and **D** (the displacement field that exists in the presence of electrically polarizable matter); and two magnetic fields: **B** (called the magnetic induction by physicists, the magnetic flux density by electrical engineers) and **H** (the magnetic field that exists in the presence of matter with a magnetic moment)—to one another, to a vector **J** representing the volume density of free electrical current flow, and to a scalar quantity ρ representing the volume density of free electric charge. In terms of these five vectors and ρ, Maxwell's equations are written in the rationalized MKSA system of units as follows, where vectors are signified by boldface type:

$$\text{div}\,\mathbf{D} = \rho \qquad \mathbf{curl}\,\mathbf{E} + \partial\mathbf{B}/\partial t = 0$$
$$\text{div}\,\mathbf{B} = 0 \qquad \mathbf{curl}\,\mathbf{H} - \partial\mathbf{D}/\partial t = \mathbf{J}.$$

The theory of relativity teaches that electric and magnetic fields are entangled: that is, when a transformation is made from one reference frame to another that is moving relative to the first one, the expression for the electric field in one reference frame depends on both the electric and magnetic fields in the other reference frame. A similar statement is true for magnetic fields.

A fact not widely recognized by those who study the health effects of exposure to non-ionizing electromagnetic fields is that both energy and momentum are associated with an electromagnetic wave. To put it in terms familiar to electrical engineers, an electromagnetic wave transmits both energy and momentum.

Energy and momentum are another pair that are similarly entangled, just as electric and magnetic fields are. What this means is that a complete evaluation of the effects of the interaction between a non-ionizing electromagnetic field and matter must consider the consequences of both energy and momentum transfer. Because of their education and training, electrical engineers have focused on the consequences of energy transfer, while disregarding the consequences of momentum transfer.

In a vacuum, the density (per unit volume) of momentum associated with an electromagnetic field is defined[3] as

$$\mathbf{g} \equiv \mathbf{D} \times \mathbf{B}$$

which means that, in a vacuum,

$$\mathbf{g} = (1/c^2)\,\mathfrak{P}.$$

This states that, in a vacuum, the Poynting vector $\mathfrak{P} \equiv \mathbf{E}\times\mathbf{H}$ which measures the electromagnetic power flux, the areal density of power flow, and the vector $\mathbf{g}$ representing the momentum density of the electromagnetic field are parallel but of vastly different magnitudes, that of the Poynting vector being larger by a factor of c^2, where c is the velocity of light in a vacuum. This is one reason why the profession of electrical engineering typically focuses on the energy-transmitting properties of the electromagnetic field and ignores its momentum-transmitting properties: The amount of energy transmitted by an electromagnetic wave is far greater than the amount of momentum transmitted.

There has been controversy throughout the twentieth century regarding the correct definition of $\mathbf{g}$ within a material medium. Minkowski[3] has asserted that the definition $\mathbf{g} \equiv \mathbf{D}\times\mathbf{B}$ is valid in all media, while Abraham[4] preferred the definition $\mathbf{g} \equiv (1/c^2)\,\mathbf{E}\times\mathbf{H}$ in all media. These two definitions become identical in a vacuum. (A discussion of the controversy at the midpoint of the twentieth century is given by Møller.[5] I assume that Minkowski's is the correct definition of $\mathbf{g}$ because I believe that the phenomenon of birefringence arises from $\mathfrak{P}$ and $\mathbf{g}$ taking different paths through a birefringent crystal, which Abraham's definition of $\mathbf{g}$ does not permit. If my assumption is correct, then in a linear isotropic homogenous dielectric medium within which υ is the speed of light, $\mathbf{g} = (1/\upsilon^2)\,\mathfrak{P}$, an equation which can be found in the advanced textbook by Portis.[6] It is not yet clear whether this relationship between g and $\mathfrak{P}$ also holds in conductive media; it would seem to be inconsistent with the definition of $\mathbf{g}$ in such media.)

Table 1 compares key parameters in different material media, displaying formulas. It makes evident that only in a conducting medium does absorption of energy and momentum take place. The equations found in electrical engineering textbooks are mostly the simpler equations for a dielectric medium, probably because

Parameter	Linear Isotropic Homogeneous Media		
	Vacuum	Perfect Dielectric	Electrically Conducting Medium
σ	0	0	$0 < \sigma < \infty$
$Q \equiv \omega\epsilon/\sigma$	∞	∞	$0 < Q < \infty$
Speed of Light	$c = \lambda_o f$	$v = \lambda f$	$v_c = \frac{1}{\sqrt{\mu\epsilon}} \left[\frac{2}{[1+(1/Q^2)]^{1/2} + 1} \right]^{1/2}$
Relation to Medium	$1/c^2 = \mu_o \epsilon_o$	$1/v^2 = \mu\epsilon$	$1/v_c^2 = \frac{1}{2} \mu\epsilon \{[1+(1/Q^2)]^{1/2} + 1\}^{1/2}$
Wavelength	$\lambda_o = 2\pi c/\omega$	$\lambda = 2\pi v/\omega$	$\lambda_c = (2\sqrt{2}\pi/\omega)/[\mu\epsilon \{[1+(1/Q^2)]^{1/2} + 1\}]^{1/2}$
Propagation Constant k	real $k^2 = \omega^2/c^2$ $\lvert k \rvert = 2\pi/\lambda_o$	real $k^2 = \omega^2/v^2 = \omega^2\mu\epsilon$ $\lvert k \rvert = 2\pi/\lambda$	complex $k_c^2 = \omega^2\mu\epsilon (1+\sqrt{-1}/Q)$ $\lvert k \rvert = \omega\sqrt{\mu\epsilon}\ [1+(1/Q^2)]^{1/4}$
Phase Angle $\Omega = \measuredangle$ **E**,**B**	$\Omega = 0$	$\Omega = 0$	$0 < \Omega < 45°$ $0 < \Omega < \pi/4$ radians $\tan\Omega = [1+Q^2]^{1/2} - Q$
amplitude **B** / amplitude **E**	$1/c$	$1/v = \sqrt{\mu\epsilon}$	$\sqrt{\mu\epsilon}\,[1+(1/Q)^2]^{1/2}$
Absorption of: energy?	NO	NO	YES
momentum?	NO	NO	YES

Table 1. The plane electromagnetic wave (**E** $\perp$ **B**) in various linear isotropic homogeneous media[7], whereby $\omega = 2\pi f$, $\mathbf{D} = \epsilon\mathbf{E}$, $\mathbf{D} \parallel \mathbf{E}$, $\mathbf{B} = \mu\mathbf{H}$, $\mathbf{B} \parallel \mathbf{H}$, ϵ is the permittivity of the medium, μ is the permeability of the medium, and σ is the electrical conductivity of the medium.

air is a dielectric medium and in many practical instances, electrical engineers are concerned only about electromagnetic fields in air.

Energy and Momentum

Two different effects are possible when non-ionizing electromagnetic radiation interacts with matter: a thermal effect and a nonthermal (or athermal) effect. Thermal health effects, which arise from the absorption of energy by matter with resultant heating, have long been accepted as real. The existence of nonthermal health effects has been controversial for many decades, with key members of the electrical engineering profession in the forefront of those skeptical of their existence, although it is now acknowledged by most scientists actively studying the biological effects of such fields that nonthermal health effects are indeed real.

The theoretical foundation for bioelectromagnetic hygiene provides the needed explanation for nonthermal health effects: They result from the absorption of momentum by matter. An electromagnetic wave impinging upon any material medium, biological or otherwise, has the potential to generate both thermal and nonthermal effects simultaneously. (As will be seen, the two kinds of health effects are not both observed under the same identical set of circumstances, at the usual field strengths encountered.)

Most students of non-ionizing electromagnetic field bioeffects understand what it means to say that matter absorbs energy from an electromagnetic field. They understand that it implies that electric currents are induced in the material medium, and the Joule heating associated with these induced currents raises the temperature of the material medium within which they flow. It is generally accepted that the damage to health associated with thermal effects results from the generation of heat in this manner within the living organism. (The specific absorption rate [SAR] is the parameter usually employed to quantitate the rate of heat generation in material media, including biological media, for the purpose of determining the thermal health hazard.)

But what does it mean to say that matter absorbs momentum from an electromagnetic field? The time rate of change of momentum is force. The absorption of momentum implies that a force is exerted.

Electric and magnetic fields can exert a force only on charges and currents, that is, on static or moving charges. The Lorentz force is the force exerted by external electric (**E**) and magnetic (**B**) fields on a particle with charge q that is moving at velocity **v**; the equation for the Lorentz force is

$$\mathbf{F} = q(\mathbf{E} + \mathbf{v}\times\mathbf{B}).$$

If this equation is rewritten so that it applies to charge density and current density—the volume density of current and charge—then the force becomes a volume density of force **f**, for which the equation is

$$\mathbf{f} = \rho\mathbf{E} + \mathbf{J}\times\mathbf{B}.$$

This equation can be expressed entirely in terms of the vectors of the electromagnetic field by making substitutions from Maxwell's equations of electromagnetism to obtain

$$\mathbf{f} = (\mathrm{div}\ \mathbf{D})\mathbf{E} + (\mathbf{curl\ H}) \times \mathbf{B} - \partial\mathbf{D}/\partial t \times \mathbf{B}$$

to which may be added the equation

$$0 = (\mathrm{div}\ \mathbf{B})\mathbf{H} - \mathbf{D} \times (\mathbf{curl\ E} + \partial\mathbf{B}/\partial t)$$

to produce

$$\mathbf{f} = (\mathrm{div}\ \mathbf{D})\mathbf{E} + (\mathrm{div}\ \mathbf{B})\mathbf{H} - \mathbf{D}\times\mathbf{curl\ E} - \mathbf{B}\times\mathbf{curl\ H} - \partial(\mathbf{D}\times\mathbf{B})/\partial t$$

which in a linear isotropic homogeneous medium may be written

$$\mathbf{f} = \mathrm{div}\ \mathbf{T} - \partial\mathbf{g}/\partial t$$

where **T** is the Maxwell stress tensor [see Appendix A] and the divergence of **T** is a vector.

In the special case of a single plane electromagnetic wave, div $\mathbf{T} = -\nabla u \equiv$ **grad** u—the gradient of u—where u, the electromagnetic energy density (per unit volume), is defined as

$$u \equiv \tfrac{1}{2}(\mathbf{D}\cdot\mathbf{E} + \mathbf{B}\cdot\mathbf{H})$$

from which it follows that

$$\mathbf{f} = -\ \mathbf{grad}\ u - \partial\mathbf{g}/\partial t$$

for this very simplest of wave shapes. In a plane electromagnetic wave, all three vectors (**f**, **grad** u, and $\partial\mathbf{g}/\partial t$) lie parallel or antiparallel to the direction of propagation of the wave.

In a vacuum or a perfect dielectric, **f**=0 which means that **grad** $u = -\ \partial\mathbf{g}/\partial t$ in these media. This makes it clear that the two vectors in the equation for **f** are antiparallel in a perfect dielectric: They point in opposite directions. In order for **f** to be nonzero, one of these vectors must have a larger magnitude than the other, or they must not lie along the same straight line, or both. It appears that the **grad** u term is the dominant one in biological media, though this has not yet been confirmed by theoretical calculations applicable to a conductive medium.

Because it is customary to utilize either the power density (the magnitude of the Poynting vector $\mathfrak{P}$) or measurements of the electric field strength $|\mathbf{E}|$ and the magnetic field strength $|\mathbf{H}|$ or $|\mathbf{B}|$ to evaluate the hazard to health posed by exposure to electromagnetic radiation, it is useful to inquire whether **f** can be expressed in terms of either of these.

The magnitude of the Poynting vector is always given by

$$|\mathfrak{P}| = |\mathbf{E}|\ |\mathbf{H}| \sin\alpha$$

where α is the smaller angle between **E** and **H**. In a plane wave, which is being considered here, this angle is always $\pi/2$, or 90°. In a linear isotropic homogeneous material medium $u = \frac{1}{2}(\epsilon E^2 + \mu H^2)$, and in a dielectric medium, $\mathfrak{P} = \upsilon^2 \mathbf{g}$, where υ is the speed of light in the material medium. In a perfect dielectric, $1/\upsilon^2 = \mu\epsilon$; in a conducting medium, υ is reduced. Even if the equation $\mathfrak{P} = \upsilon^2\,\mathbf{g}$ fails to hold in a conductive medium, it is nevertheless highly likely that **g** remains proportional to $\mathfrak{P}$ in any medium that is not strongly anisotropic, in which case the relation

$$\mathbf{g} \propto |\mathbf{E}|\ |\mathbf{H}| \sin\alpha$$

probably holds in soft biological tissues such as blood, muscle, skin, brain, and other internal organs.

There is no way that **grad** u can be expressed as a function of $|\mathfrak{P}|$, but because **grad** u is the sum of two similar functions, one of the square of the electric field strength, and the other of the square of the magnetic field strength, **grad** u will become larger as $|\mathfrak{P}|$ increases. The same is true of $\partial\mathfrak{P}/\partial t$ and therefore of $\partial\mathbf{g}/\partial t$. So $|\mathbf{f}|$ will indeed increase as $|\mathfrak{P}|$ increases. (This suggests that it may be possible to base a practical exposure standard providing partial protection against the nonthermal health effects of exposure to a microwave or radio-frequency electromagnetic field on the power density of the radiation, just as is currently done to protect against the thermal health hazard of such exposure, though such a standard will surely be quite different from current standards.)

Clearly, then, for a plane electromagnetic wave in a linear isotropic homogenous dielectric medium, such as a vacuum or air, **f** can be expressed as a function of the electric and magnetic field strengths, though not specifically as a function of the power density $|\mathfrak{P}|$:

$$\mathbf{f} = -\tfrac{1}{2}\,\epsilon\,\mathbf{grad}\,E^2 - \tfrac{1}{2}\,\mu\,\mathbf{grad}\,H^2 - (1/\upsilon^2)\,\partial(\mathbf{E}\times\mathbf{H})/\partial t$$

$$= -\,\epsilon E\,\mathbf{grad}\,E - \mu H\,\mathbf{grad}\,H - \epsilon\mu\,[\mathbf{E}\times\partial\mathbf{H}/\partial t + \partial\mathbf{E}/\partial t\times\mathbf{H}].$$

Because all the terms on the right are vectors, their vector sum cannot be represented any more simply than this. However, it is possible to express this in terms of the other pair of electric and magnetic vectors, **D** and **B**:

$$= -\,(D/\epsilon)\,\mathbf{grad}\,D - (B/\mu)\,\mathbf{grad}\,B - [\mathbf{D}\times\partial\mathbf{B}/\partial t + \partial\mathbf{D}/\partial t\times\mathbf{B}]$$

which may be considered to simplify the expression slightly.

For electromagnetic waves other than plane waves, there are additional terms in the equation, so that

$$\mathbf{f} = \mathrm{div}\,\mathbf{DE} + \mathrm{div}\,\mathbf{BH} - \mathbf{grad}\,u - \partial\mathbf{g}/\partial t.$$

Where there are multiple radiation sources in operation, or where there is a single source of a non-plane wave, this is the equation for **f** that applies.

It should be emphasized that these equations, strictly speaking, apply only to linear isotropic homogeneous material media, and therefore cannot be assumed to apply accurately to biological media, which are typically nonlinear, anisotropic, and nonhomogeneous as well as elastic (non-rigid) and within which many different electrochemical reactions occur in order to sustain life. Nevertheless, these equations do provide an approximation to what occurs in biological media, so they provide useful guidance that can assist in gaining an understanding of the concepts they embody. They also point quite clearly toward testable exposure metrics for nonthermal health hazards for which a theoretical foundation exists.

An argument can be made that cancer, which has been associated with exposure to various types of non-ionizing electromagnetic fields, must be a nonthermal health effect, because heat is used to kill cancer cells, therefore a thermal effect is unlikely to cause cancer cells to thrive and grow in an unrestrained manner. It is no longer necessary to rely on this argument, however, because there is now experimental evidence showing that cancer resulting from exposure to non-ionizing electromagnetic radiation is indeed a nonthermal health effect.[8]

The Cancer Question

Does exposure to microwave radiation cause cancer? There are data from both human experience and controlled laboratory studies of animals indicating that it is certainly a risk factor in the development of cancer. Leaving aside for the moment the human epidemiological evidence, which in any case shows only association, not causation, the best evidence of a cancer hazard to mammals resulting from long-term, low-level exposure to pulsed microwave radiation comes from two different long-term controlled laboratory studies of specific-pathogen-free rodents. The first study[9] was of rats; its results had become available by 1986, although the complete scientific report of the results of this study was not published for another 6 years.

This experiment was carried out on 200 male rats that were born into a sterile environment and maintained there throughout their lives. Of these, 100 were exposed inside a waveguide to pulsed, square-wave-modulated circularly polarized 2450 MHz radiation for 21.5 hours per day for 25 months, beginning at 8 weeks of age, while the other 100 were sham-exposed. The average SAR for an exposed rat weighing 200 grams was 0.4 W/kg; at the end of the experiment, for rats weighing 800 grams, the SAR would have been 0.15 W/kg. The data on primary malignant tumors, without regard to the tissues or organs within which they developed, were as follows: Eighteen of the 100 irradiated rats, but only 5 of the 100 sham-irradiated rats, developed a primary malignancy somewhere in their body during their lifetime.

I used Fisher's exact test to find the significance level of these totals, employing them as entries in a 2×2 contingency table; I obtained $p \approx 0.0020333$ for these data. (For the significance level it produces, Fisher's exact test is the uniformly most powerful unbiased (UMPU) statistical test that can be applied to these kind of data.[10] Therefore, as there does not exist a UMP test for such data, I consider Fisher's exact test the best statistical test to use, simply because it is always best to use the most powerful statistical test possible.)

A more recent study[11] was carried out in Australia on transgenic mice, also born into and maintained in a sterile environment. Because the purpose of this study was to evaluate the health hazards of exposure to microwave radiation from mobile telephones, the mice were exposed to 900 MHz pulsed radiation from a quarter-wave monopole antenna in the center of a ground plane (which simulates the radiation from a half-wave antenna in a digital hand-held mobile phone). This controlled experiment lasted for a year and a half. Again, the microwave-irradiated mice displayed a higher incidence of lymphomas, the malignancies they were genetically predisposed to develop, when compared to the control group. The authors obtained a statistically significant odds ratio, for which they found $p<0.006$. Had they tested their data using Fisher's exact test, they would probably have obtained an even lower value for p.

The results from both rodent experiments were highly statistically significant, meaning they were quite unlikely to have occurred by chance, in the absence of a real effect. Ideally, of course, these experiments should be replicated. The rat study was sponsored by the US Air Force. But in the past 15 years, no agency of the U.S. or any other government has shown any interest at all in replicating this experiment on rats or in carrying out any other similar study. (The sponsor of the study of transgenic mice was Telstra Corporation Limited, a wireless telecommunications firm in Australia.)

These two controlled lifetime laboratory studies of rodents have produced results showing that there is a cancer hazard to rodents—and presumably to other mammals, as well—associated with whole-body exposure to microwave radiation. Because these were controlled studies, they have shown quite clearly that the association is with exposure to microwave radiation, and not with exposure to some other environmental agent. While these two studies were not exact replications, they did test the same basic hypothesis in substantially the same way, obtaining very similar results.

How likely is it that two such studies would produce the same kind of result, if there were no real effect of chronic low-level microwave radiation on cancer incidence? This probability can be estimated simply by multiplying the p-values of the data in the two studies; the product is ≈ 0.0000122, a bit larger than 1 chance in 100,000. This value is so low that the null hypothesis of no effect may be rejected with enormous confidence, and the conclusion may be drawn that microwave radiation can indeed increase the incidence of cancer in rodents subjected to whole-body microwave irradiation at quite low radiation intensities. If these results are characteristic of mammals in general, as seems likely, then the implication is that pulsed microwave radiation at these two frequencies—900 and 2450 MHz—is capable of causing cancer in human beings who are chronically exposed to it.

This, in conjunction with strongly suggestive data on human populations, most of which comes from epidemiological data, indicates the existence of a cancer hazard to mammals associated with exposure to microwave radiation at 900 MHz and 2450 MHz, and possibly also at other frequencies between these two values.

Carcinogenicity is Associated with Very Low Microwave Radiation Intensity

The most strongly suggestive evidence in a human population of a cancer hazard from exposure to microwave radiation comes from the experience of law enforcement officers who used the earliest models of traffic radar guns. Furthermore, this evidence makes abundantly clear that this cancer hazard lies well below the alleged "safe limit" of 10 mW/cm² established by ANSI C95.

Traffic radar guns are used to measure vehicle speed on roads and highways. These devices were manufactured to comply with ANSI C95, and they did so within a considerable margin of safety. In tests of a variety of different types extending over a decade,[12] the average aperture power density ranged from 0.28 to 3.36 mW/cm². In a study of operator exposure, much of the time the power density to which the operator was exposed was less than 1% of the aperture power density of the instrument.[12] The power densities to which the users of these devices were subjected were well below the ANSI safe limit of 10 mW/cm², usually between 2.8 and 33.6 μW/cm², with some excursions to somewhat higher levels. If ANSI C95 actually protected against both thermal and nonthermal health hazards, as electrical engineers today assert it does, then these traffic radar guns should have been perfectly safe for law enforcement officers to use.

Law enforcement officers developed cancer at or very close to body sites that had been chronically irradiated by early-model traffic radar guns, which were designed to emit a continuous beam of microwave radiation that could not be turned off and on without recalibration of the instrument. As a result, officers had to do something with the instrument when it was not actually being used to measure the speed of a vehicle, and because they did not want to alert radar detectors in passing cars, they often tried to "bury" the beam. Different officers used different methods, but a particular method tended to become habitual for each individual officer. As a result, if the method allowed body tissues to be exposed, those same body tissues tended to be exposed over and over again, day after day, week after week, month after month, creating a condition of chronic low-level exposure to microwave radiation.

Male officers who habitually laid their traffic radar gun on their lap reported developing testicular cancer.[13] A motorcycle officer slung his over his shoulder, so that the beam grazed his back; he subsequently reported developing cancer in the tissues of his back. Either the edge of the main beam, or a side lobe from the main beam, would have irradiated the body tissues where cancer developed, meaning that if the cancers reported in these law enforcement officers were truly associated with the chronic low-level exposure to microwave radiation they reported receiving from their traffic radar guns, the irradiated tissue where the cancer developed would have been exposed to microwave radiation at an intensity substantially below the level recommended by ANSI C95.

A US Senate subcommittee hearing was held on this issue in August 1992, chaired by Senator Joseph I. Lieberman (D-CT),[14] at which it was reported that the number of law enforcement officers so affected was at least 168, all of them males except for one female. Most of these officers died of their cancers.

The experience of these law enforcement officers is a clear indication that, under conditions of chronic irradiation, very low-intensity microwave radiation is a risk factor in the subsequent development of cancer in the tissue so irradiated—provided one can be sure that this cluster of cancer cases was not a random statistical cluster.

I attended the hearing chaired by Senator Lieberman, heard the testimony, and afterward carried out a rough statistical test of the random statistical cluster hypothesis, using a binomial probability distribution and assuming that the total number of law enforcement officers (police and state troopers) who had used these early-model traffic radar guns for an extended period of time in the U.S. was 600,000, in order to test the count of 168 officers with cancers alleged to be associated with traffic radar gun microwave exposure. The association of their cancers with their previous microwave exposure had been made by the affected officers because of the location on their body where their cancer appeared. I therefore assumed that all 168 had been fated to develop cancer somewhere in their body at the time it actually developed as the result of some unidentified other influence—the null hypothesis—and calculated the likelihood that such cancers would have developed, quite by chance, at the very body site that had previously received chronic microwave irradiation. I obtained $p<.000001$, an exceedingly low value—so low that if it had been obtained in a clinical medical trial, it would have brought the trial to an immediate halt.

Just to be on the safe side, I tested my model for sensitivity to assumed values of parameters and found that these values could be varied greatly without affecting the conclusion. This gave me the confidence to reject the null hypothesis and accept these 168 cancer cases as genuinely (and probably causally) associated with the officers' previous chronic microwave irradiation associated with use of their traffic radar gun.

Notice that this provides confirmation of the results obtained from laboratory studies of rodents in a different mammalian species, *homo sapiens*, and indicates that the laboratory findings on rodents are probably generalizable to other mammalian species. For skeptics, an estimate of the probability that the two laboratory studies of rodents and the experience of law enforcement officers using early model traffic radar guns all three gave misleading results, quite by chance, in the absence of a real effect, can be obtained by multiplying the three *p*-values: $0.0020333 \times 0.006 \times 0.000001 \approx 0.0000000000122$, which is an absurdly small value, so close to zero that for all practical purposes one may declare the question of the carcinogenicity of microwave radiation permanently settled, the only remaining question being to determine, not whether microwave radiation is carcinogenic, but under what conditions this carcinogenicity manifests itself.

Based on the totality of the available evidence, especially the experience of the law enforcement officers who used the early-model traffic radar guns, there is considerable evidence of a cancer hazard associated with microwave exposure; and this hazard is much larger at radiation intensities well below the 10 mW/cm^2 level of the ANSI standard than it is at intensities near or above this value. Therefore, as the microwave radiation intensity drops below 10 mW/cm^2, the cancer hazard apparently increases. And because cancer is a nonthermal health hazard, it is reasonable to suppose that the risk of all nonthermal health hazards increases as the radiation intensity decreases below 10 mW/cm^2.

The Challenge of the Exposure Metric

Thus far, it has been shown that nonthermal health hazards resulting from the absorption of electromagnetic momentum within living tissue can be expected to exist, and a serious human health problem associated with exposure to microwave radiation has been identified as such: cancer in law enforcement officers who used early-model traffic radar guns on the job.

These people were exposed to microwave radiation which produced these effects because the equipment they used was declared safe for human use. This happened because the human exposures that resulted were considered to be safe, based on a health standard the content of which was almost exclusively determined by the electrical engineering profession.

Existing health protection standards are based on thermal hazards. Now that it is clear that there is a nonthermal hazard to human health from exposure to microwave radiation, what are the implications with respect to the quantitation of human exposure? Specifically, what exposure metric should be used?

SAR has long been used by electrical engineers as a measure of exposure when the thermal effects upon human health of exposure to microwave radiation are being quantitated. Table 2 shows that the bioelectromagnetic hygienist may reasonably be asked, "Is there anything comparable to the SAR that can be used as a measure of exposure to nonthermal health risks?"

At the present time it is not known whether any parameter comparable to the SAR exists for nonthermal health hazards; though if any such parameter does exist, it is likely to be a vector quantity, not a scalar quantity. The attempt to answer this question is one of the major challenges currently facing the profession of bioelectromagnetic hygiene.

Thermal health hazards	Nonthermal health hazards
energy absorption by matter	momentum absorption by matter
SAR	?
(scalar)	(vector)

Table 2. Comparable exposure parameters and mechanisms.

The Need for Two Exposure Metrics

Table 2 suggests that a single exposure metric may not be sufficient to provide protection against all the adverse health effects of exposure to microwave and radio-frequency electromagnetic radiation. Two different exposure metrics would appear to be needed: one to protect against thermal health effects and another to protect against nonthermal health effects.

Because the idea that two different exposure metrics are needed to quantitate the health risks of exposure to a non-ionizing electromagnetic field may seem strange, it is appropriate to draw an analogy with the health hazards posed by the chemical pesticide lead arsenate, Pb_3AsO_4. This was commonly used as an insecticide for fruit in orchards until the past few decades.

Its chemical composition shows that it contains arsenic (As) and lead (Pb), both of which are toxic. Exposure to lead arsenate therefore presents two separate hazards to health, one from lead and one from arsenic. The lead poisoning hazard it poses can be assessed by measuring the exposure to lead it produces; the arsenic poisoning hazard it poses can be determined by measuring the exposure to arsenic it produces. Notice that the lead poisoning hazard is not determined by measuring arsenic exposure, nor is the arsenic poisoning hazard evaluated by measuring lead exposure. One must measure the exposure that is appropriate to the particular health hazard of interest.

The same is true for non-ionizing electromagnetic radiation, which poses both thermal and nonthermal health hazards to biological organisms. One cannot measure the SAR and then expect that this provides an appropriate basis for assessing the risk to health from both thermal and nonthermal effects. The SAR is appropriate only for determining the thermal hazard. Something else must be employed to evaluate the nonthermal health hazard.

Determining what exposure metric is best for evaluating nonthermal health risks is a central concern of bioelectromagnetic hygiene at this time.

The Basis for a Thermal Exposure Standard

For comparison, the thermal hazard metric comparable to **f** is derived from the well-known energy balance equation, Poynting's theorem, which in integral form is

$$\int_{V} \mathbf{J}\cdot\mathbf{E}\, d\tau = -\int_{V} (\mathbf{E}\cdot\partial\mathbf{D}/\partial t + \mathbf{H}\cdot\partial\mathbf{B}/\partial t)\, d\tau - \oint_{S} \mathfrak{P}\hat{\mathbf{n}}\, dS$$

where S is the surface of volume V and $\hat{\mathbf{n}}$ is an outward-pointing unit vector normal to this surface. Transforming the surface integral to a volume integral gives

$$\int_{V} \mathbf{J}\cdot\mathbf{E}\, d\tau = -\int_{V} (\mathbf{E}\cdot\partial\mathbf{D}/\partial t + \mathbf{H}\cdot\partial\mathbf{B}/\partial t)\, d\tau - \int_{V} \operatorname{div} \mathfrak{P}\, d\tau$$

and because the volume is arbitrary, the integrands may be equated to give an equation valid in any material medium

$$\mathbf{J}\cdot\mathbf{E} = -(\mathbf{E}\cdot\partial\mathbf{D}/\partial t + \mathbf{H}\cdot\partial\mathbf{B}/\partial t) - \operatorname{div} \mathfrak{P} \quad \text{[any material medium]}$$

which in the special case of a linear isotropic homogeneous medium becomes

$$\mathbf{J}\cdot\mathbf{E} = -\partial u/\partial t - \operatorname{div} \mathcal{P} \quad \text{[linear isotropic homogeneous medium]}$$

which is easier to conceptualize.

In a mechanically rigid medium (where the transformation of electromagnetic energy into the elastic energy of a stressed medium cannot occur) which is free of hysteresis effects and in which no impressed electromotive force is causing a flow of current against existing forces,[15] $\mathbf{J}\cdot\mathbf{E}$ is the rate (per unit volume) at which electromagnetic energy is dissipated as heat, which will be termed w; thus $w = \mathbf{J}\cdot\mathbf{E}$ in such a medium.

Table 3 illustrates that w and $\mathbf{f}$ are corresponding local metrics for thermal and nonthermal effects in material media, respectively. These are conceptually valuable metrics, but not useful in a practical sense. To get practical metrics, the volume integrals of these must be examined.

Thermal health hazards	**Nonthermal health hazards**
rate of heat generation/unit volume	rate of momentum generation/unit volume
w	$\mathbf{f}$

Table 3. Local metrics.

An Integral Form for the Exposure Metrics

To obtain from w a practical metric for thermal hazard in a linear isotropic homogeneous medium, the original integral equation expressing Poynting's theorem is used. The rate of heat generation within a volume V is given by the following volume integral:

$$\mathrm{W} = \int_{V} w \, d\tau$$

where $d\tau$ is the infinitesimal element of volume. The time-average $\overline{\mathrm{W}}$ of W is, by definition,

$$\overline{\mathrm{W}} \equiv (1/\mathrm{T}) \int_{0}^{\mathrm{T}} \mathrm{W} \, dt$$

where $\mathrm{T} = 2\pi/\omega = 1/f$ is the period of the oscillation. Thus

$$\overline{W} \equiv (1/T) \int_0^T \left[\int_V w \, d\tau \right] dt$$

which can be evaluated when the expression for w (as a function of the three spatial variables and the time) is known. In performing the calculation, of course, the order of integration can be interchanged.

In the media considered here, $w = \mathbf{J} \cdot \mathbf{E}$, hence

$$\overline{W} \equiv (1/T) \int_0^T \left\{ \int_V [- \operatorname{div} \boldsymbol{\mathcal{P}} - \partial u/\partial t \,] \, d\tau \right\} dt.$$

The time integral of the second term of the integrand is zero:

$$\int_0^T \partial u/\partial t \; dt = [\, u \,]_0^T = [u(T) - u(0)] = 0$$

so only the first term of the integrand, $- \operatorname{div} \boldsymbol{\mathcal{P}}$, contributes to the expression for $\overline{W}$.

Where a plane wave approximation is valid, Cartesian coordinates are appropriate and the volume integration is performed over a cube of material. For convenience, it will be assumed that the radiation is traveling in the direction of the positive z axis, so that $\boldsymbol{\mathcal{P}} = \mathcal{P}_z \, \hat{\mathbf{z}}$, and $\mathcal{P}_x = \mathcal{P}_y = 0$. Because this is a plane wave, $\boldsymbol{\mathcal{P}}$ is a function only of z and t, not of x or y.

$$W = \int_V (- \operatorname{div} \boldsymbol{\mathcal{P}}) \, d\tau$$

$$= - \int\int \left[\int_{x_1}^{x_2} \partial \mathcal{P}_x/\partial x \, dx \right] dy dz - \int\int \left[\int_{y_1}^{y_2} \partial \mathcal{P}_y/\partial y \, dy \right] dx dz - \int\int \left[\int_{z_i}^{z_f} \partial \mathcal{P}_z/\partial z \, dz \right] dx dy$$

$$= \left[\int_{z_f}^{z_i} \partial \mathcal{P}_z/\partial z \, dz \right] \left[\int_{y_2}^{y_1} \int_{x_2}^{x_1} dx dy \right] = [\mathcal{P}_z\,]_{z_f}^{z_i} \, A = [\mathcal{P}_z(z_i) - \mathcal{P}_z(z_f)] \, A$$

where $A = (x_1 - x_2)(y_1 - y_2)$ is the area of the xy face of the cube, $\mathcal{P}_z(z_i)$ is the power density incident on this cube face, and $\mathcal{P}_z(z_f)$ is the power density of the radiation exiting the volume through the opposite xy face of the cube.

In a worst case scenario, all incident radiation will be absorbed and none will exit the cubical volume; that is, $\mathcal{P}_z(z_f)$ will be zero. In that case,

$$W = \mathcal{P}_z(z_i) \, A = |\boldsymbol{\mathcal{P}}| \, A \quad \text{(all incident plane-wave radiation is absorbed)}$$

so the rate at which power is dissipated as heat in this volume will then be the product of the magnitude of the Poynting vector (the power density) and the area irradiated. (The time-average $\overline{W}$ of W is then calculated from the time-average of the magnitude of the Poynting vector.)

This derivation demonstrates the scientific basis for establishing an upper limit on the value of the power density as a means by which to control the thermal hazard to health. The rate at which heat is generated within the volume V is proportional to the power density of the incident radiation, if all or most of it is absorbed; then to keep W from getting too high, an upper limit must be imposed upon $|\mathfrak{P}|$. This is what ANSI C95 does—and what all other existing health standards for human exposure to non-ionizing radiation also do.

However, it should be noticed that for this relation to be valid, the radiation must be approximated by a plane wave, and absorption should be virtually 100%. This equation is valid only for the rate of heat generation within the volume V, so any standard derived from it can protect only against a thermal hazard to health. (This is currently a point of contention;[16] see Appendix B for a discussion.)

Can the same thing be done for the nonthermal health hazard? In principle, yes; in practice, not yet. The applicable equation for a linear isotropic homogeneous medium is

$$\mathbf{f} = \mathrm{div}\,\mathbf{T} - \partial\mathbf{g}/\partial t = \mathrm{div}\,\mathbf{DE} + \mathrm{div}\,\mathbf{BH} - \mathbf{grad}\,u - \partial\mathbf{g}/\partial t$$

where the problem is that a correct expression for $\mathbf{g}$ in a conductive medium is not yet known with certainty. In a perfect dielectric medium, where the expression for $\mathbf{g}$ *is* known, there is no absorption of either energy or momentum, as Table 1 shows, so the resulting equations can represent nothing quantitative that has meaning.

One course of action is to assume that $\partial\mathbf{g}/\partial t$ is so small in magnitude that it can be neglected. For sufficiently low frequencies, this is reasonable, because the magnitude of $\mathbf{g}$ is a very small fraction of the magnitude of the Poynting vector. Then

$$\mathbf{f} \simeq \mathrm{div}\,\mathbf{T} \quad \text{(low frequency approximation)}$$

and the volume integral of this is the total force $\mathbf{F}$ exerted by fields external to V on the charges and currents within V. Because this force must pass through the surface enclosing this volume, the total force acting in the interior of V can be represented by surface forces, surface stresses, acting at the bounding surface S, which has an outward-pointing normal unit vector $\hat{\mathbf{n}}$. If $\mathbf{f}$ can be expressed as the divergence of a tensor, then the component of that tensor normal to the surface defines the surface stress $\mathbf{t}$ at the point of interest:[17]

$$\mathbf{t} \equiv \mathbf{T} \cdot \hat{\mathbf{n}}$$

and the volume integral of the divergence of $\mathbf{T}$ can be transformed to a surface integral of $\mathbf{t}$:

$$\mathbf{F} = \int_V \mathbf{f}\, d\tau = \iiint_V \operatorname{div} \mathbf{T}\, d\tau = \oint_S \mathbf{t}\cdot\hat{\mathbf{n}}\, dS.$$

Notice that **t**, the surface force or surface stress, may be tensile, compressive, or shear.[18] (This topic is discussed at length in the literature on electromagnetism from the late nineteenth and early twentieth centuries.)

Tensile and compressive stresses are pressures. A compressive stress is a pressure directed into the volume, normal to the surface; a tensile stress is a pressure directed out of the volume, again normal to the surface. If these forces arise from an electric field **E** only, and act on the surface of a cube, the surface stress **t** will be tensile where **E** is normal to a cube face, and compressive where **E** lies in the surface of a cube face.

A more realistic shape is that of a sphere, which approximates the shape of some biological cells, such as ova. If the angle between the surface normal and **E** is neither zero nor a multiple of 45°, the surface stress **t** will be a combination of a shear stress and a normal stress (either compressive or tensile). If this angle is exactly 45° or 135° the surface force will be a pure shear stress, meaning that **t** lies in the plane tangent to the surface. (These statements are also true if a magnetic field is substituted for the electric field.)

The theory clearly shows that the pressure inside a material medium can be altered by application of either an electric or a magnetic field. Thus it is hardly surprising that not long ago, a report was published in the medical literature[19] that the resting blood pressure in human volunteers was increased when they were exposed to a radio-frequency electromagnetic field.

The application of this phenomenon to a biological situation, where the protoplasm of a cell is enclosed by a cell membrane, is obvious. Another application would be to the breaching of the blood-brain barrier by microwave irradiation. Presumably shear forces could loosen or tear apart the closely adherent cells of the wall of a blood vessel, allowing large molecules to get through. If the forces were not too great, so that there were only tiny leaks, then when the electromagnetic field was removed, the walls of the blood vessels would probably heal themselves, demonstrating reversibility of the blood-brain barrier breach. But it is clear that if the forces were too large, or were applied for too long a time, a hematoma could result, or perhaps a permanent weakening of the blood vessel wall, eventually resulting in an aneurysm.

A subdural hematoma and a case of astrocytoma did occur in two employees who seem to have been occupationally exposed to low-intensity microwave radiation while they were working on a classified government contract at the Philco-Ford Corporation's plant in Philadelphia, PA; these ill effects became apparent in the spring of 1971.[20] I learned of these cases by talking to the man who had been medical director of that plant at that time.

A subsequent industrial hygiene investigation by the US Army's Environmental Hygiene Agency of that part of the Philco-Ford plant where these men worked "found no microwave radiation in excess of the permissible standard where employees worked."[21] The official Department of Defense triservice standard was 10

mW/cm^2, the same value as in ANSI C95.

The quoted statement shows that the Philco-Ford employees did experience microwave exposure—otherwise, the Army hygienists would have made no measurements of microwave radiation—and it also shows that the intensity level of this microwave radiation was below 10 mW/cm^2, because the industrial hygienists who evaluated the working environment judged it to be safe with respect to microwave radiation, because the radiation levels they measured were below the triservice standard of 10 mW/cm^2.[20] A case of astrocytoma and a case of subdural hematoma occurred in two engineers out of 23 who were working on the defense project at the Philco-Ford plant[21], and it seems that they were both exposed to low-intensity microwave radiation.

How far below 10 mW/cm^2 the level of microwave radiation was in the Philco-Ford plant has not been publicly reported. This information would be contained in the written report prepared by the team of industrial hygienists from the US Army's Environmental Hygiene Agency. The Department of Defense has never made this document public.

Even so, the publicly available information is important for two reasons: it shows that a case of brain cancer (astrocytoma) occurred under conditions of low-intensity exposure to microwave radiation in 1971, nearly 30 years ago, and it shows that under conditions similar to that which gave rise to the astrocytoma, a subdural hematoma occurred, providing suggestive support for the idea that bleeding within the brain may be another nonthermal effect of exposure to low-intensity microwave radiation.

An Application: The Geometry of Nonthermal Hazard

While the theory has not yet been sufficiently developed to permit identification of a useful, practical metric for regulating the nonthermal health hazard posed by exposure to a non-ionizing electromagnetic field, qualitative consideration of the momentum balance equation does enable certain useful steps to be taken to address practical problems. One such problem is the issue of nonthermal health hazards (such as brain cancer) posed by the use of portable handheld radio-telephones with an antenna in the handset, which are often called "cellular telephones" in North America, or "mobile phones" in Europe, Great Britain, South Africa, Australia and New Zealand. (Other health effects reported from the use of handheld radio-telephones with the antenna in the handset are headaches[22-23] and in some cases, impairment of mental function, including memory.[24])

In use, these radio-telephones position a microwave transmitter beside the user's head, above and behind one ear. It has been alleged—though never proven—that the microwave field from this transmitter can cause a malignant tumor to form in the brain, above and behind the ear on the side of the head where the radio-telephone is customarily used.

All transmitters are surrounded by an electromagnetic field characteristic of an efficient radio antenna. This field possesses distinctly different characteristics in different spatial regions around the antenna. Electrical engineers call the three regions the antenna region (which is immediately adjacent to the antenna), the near field, and the far field.[25]

The near field consists of the space around an antenna outside the antenna region out to a distance given by a formula that depends on the nature of the transmitter. The far field consists of the space beyond this distance. When the language of optics is used, the far field is called the Fraunhofer region and the near field is called the Fresnel region.

What the equation for **f**, the volume density of force, shows is that nonthermal health hazards depend not only on the frequency of the radiation, but also on the way the fields change spatially—that is, as one moves from point to point in space. This comes from the div **T** term.

In the case of a plane electromagnetic wave, which is a reasonable representation of a monochromatic electromagnetic wave in the far field of a source, the spatial variation is especially simple: div **T** consists of only one term, the gradient of the energy density of the electromagnetic field at the point of interest. The higher this gradient, the larger the absolute value of div **T**—and of **f**, if $\partial \mathbf{g}/\partial t$ is comparatively small in magnitude.

For an electromagnetic field or wave that is not plane, or when there is more than one plane electromagnetic wave present, there are in general three terms that contribute to div **T**, hence to the spatial variation of **f**. One of these three is the gradient of the electromagnetic energy density, but the other two do not have a simple, familiar interpretation.

Without carrying out a detailed mathematical analysis, it is evident that the spatial variation of **f** will be high in those spatial regions where the strength of the fields varies greatly over short distances. In the vicinity of a radio or microwave transmitter, this will be in the near field (Fresnel region). Thus the spatial region around a transmitter occupied by the near field is the region where the nonthermal health hazard is likely to be greatest, and the practical need is to define this region of space, that is, to locate the boundary between the near and far fields for the type of antenna being used. This boundary is located at a distance R from the antenna, and it depends on the wavelength—hence on the frequency—of the radiation.

For a thin linear antenna, the accepted equation is $R=2L^2/\lambda$ where L is the length of the antenna.[25] For aperture antennas, the boundary between near and far fields is a little more fuzzy, and either equation $R=\frac{1}{2}D^2/\lambda$ or $R=\frac{1}{4}D^2/\lambda$ may be used.[26] D is the largest linear dimension of the antenna aperture: its diameter, if the aperture is circular, or its diagonal, if the aperture is rectangular.

Of course, these equations have meaning only in reference to a specific frequency, which implies a sinusoidal waveform. A given point in space not far from a radio-frequency transmitter that is emitting radiation simultaneously at different frequencies may be in the far field of one emitted frequency and in the near field of a higher emitted frequency at the same time, because the near field of a high frequency extends farther out from the transmitter than the near field of a lower frequency does.

For a waveform that greatly departs from a sinusoidal shape, Fourier decomposition of the waveform into its component sinusoidal frequencies may be

Antenna length (cm)	Distance (m)	
	915 MHz	1900 MHz
6	0.0221 m	0.0456 m
8	0.0391 m	0.0811 m
10	0.0610 m	0.1268 m
12	0.0879 m	0.1825 m
14	0.1196 m	0.2484 m

Table 4. For a linear antenna: Distance (R) in meters in air to the near/far zone boundary, where $R=2L^2/\lambda=2L^2f/c$ and $c \approx 299{,}800$ km/sec.

employed; then the antenna may be considered to be radiating at all component frequencies simultaneously. Any calculation of the distance to the boundary between the near and far fields will be applicable only to a single component frequency.

For a transmitter of monochromatic radiation, it is a simple matter to calculate how far away from the antenna the boundary between the near and far field lies. Handheld radio-telephones do not transmit monochromatic radiation, but the analog type emits radiation over a band of frequencies that is roughly centered on the nominal frequency of operation and does not extend exceedingly far above or below it. Using the nominal frequency for calculation of the distance to the boundary between the near and far fields gives an approximate result; results applicable to mobile telephones with a linear antenna are displayed in Table 4.

For a handheld analog radio-telephone with a linear antenna 9 cm in length operating at a nominal frequency of 915 MHz, the distance to the near/far field boundary is found from Table 4 to be 5 cm, or about 2 inches. For a 13-cm antenna, the distance to the near/far field boundary would be about 4 inches.

When the relative positions of the antenna of a handheld radio-telephone and the user's head are examined, it becomes apparent that the region of highest nonthermal hazard, based on an elevated value of **f** using the semi-quantitative approach described here, intersects the head of the radio-telephone user above and behind the ear on the side of the head where the mobile telephone is used. The brain underneath this region lies in the near field of the mobile phone antenna.

Since cancer is a nonthermal health hazard, a finding of cancer in this region of the brain of a user of an analog cellular telephone with a nominal frequency of approximately 915 MHz who has a history of long or heavy use of this device prior to the appearance of the malignancy, and who typically used it on the side of the head where the cancer appeared, could be regarded as experimental confirmation of the theoretical evaluation presented here.

The approach described here is not so useful with a digital mobile telephone, because the pulsed character of its signal means that the waveform is far from sinusoidal; therefore Fourier decomposition must be employed to decompose it into its component sinusoidal frequencies, before any meaningful calculation of the near/far field boundary can be made. With respect to each of these component frequencies, the calculation can be made, because each of them has a well-defined frequency.

If this procedure were done for an analog cellular phone, the resulting distances would be clustered fairly close together, so it is convenient just to use the nominal frequency. But with digital transmitters, the component frequencies are spread over so wide a range that the calculated distances are spread over a wide range also. The result is that the nonthermal health risk is not so strongly spatially localized in the vicinity of a digital transmitter as it is in the vicinity of an analog transmitter.

So far as a hazard to the brain of the user of a digital mobile phone is concerned, the entire brain is likely to be affected. This is in contrast to analog mobile phones, which localize the nonthermal health hazard to one specific portion of the human brain, as already described.

Because of these differences in the expected health effect from use of analog and digital mobile telephones, I have advised[27] that epidemiological studies not rely on the SAR, as has been the usual recommendation[28] (and which is inappropriate in any case for a nonthermal health hazard), but instead categorize exposures according to whether they arose from use of a digital or an analog mobile telephone. Then, whether the SAR or some other metric is employed for quantification of exposure, the results of an epidemiological study are more likely to be meaningful, interpretable, and reproducible.

When considering what manifestation of adverse brain function is likely when the entire brain of a human being is exposed to a nonthermal hazard, interference with memory is a likely candidate, because memory is known not to be localized at any single site in the brain. An analog mobile phone which produces only a localized region of nonthermal hazard in the brain of the user seems unlikely to produce a memory deficit, but a digital mobile phone which spreads the region of nonthermal health hazard over the user's entire brain might well do so.

Indeed, short-term memory loss in servicemen accidentally exposed to aircraft radar has been reported in the medical literature.[29-30] Because radar is pulsed microwave radiation, it is certainly logical to assume that pulsed microwave radiation from mobile telephone handsets held close to the head could have the same effect upon the human brain.

To date, one paper has been published[31] reporting a controlled laboratory study of tests of cognitive function in human volunteers who were exposed to microwave radiation simulating that from a mobile telephone. An attempt was made in this investigation to test the short-term memory of these volunteers, with no effect reported in the published paper.[31]

Curiously, the actual experimental data from this study have not been published.[31] Only the results of a statistical analysis of the data were published; this analysis employed an unusual and possibly inappropriate statistical test, instead of the usual traditional statistical test.

What is especially curious is that the popular press in England, where the study was done, carried stories reporting an adverse effect upon short-term memory of those exposed to the simulated mobile phone microwave radiation from this study well in advance of publication of the scientific paper.[32-33] It has never been explained how the popular press apparently learned of the findings of this experiment many months before a report of the study was formally published, or why the report of the study findings published in the popular press flatly contradicted the published conclusion about memory effects contained in the scientific report, when this finally appeared in print.

Indeed, since published scientific conclusions are supposed to be accompanied by the data upon which they are based, it is surprising that any reputable scientific journal would accept for publication a report that omitted the crucial component that distinguishes a scientific report from other types of written work.

At present there are no published data answering the question of what effect exposure to mobile telephone radiation has upon the memory of the user. There are only the reports in the medical literature indicating that functional loss of human short-term memory has occurred in the past, following accidental exposure to pulsed microwave radiation. Such loss of short-term memory lasted for many weeks; indeed, the medical literature does not indicate when, if ever, the affected men recovered their short-term memory.

Pertinent to these questions is the experience of Steve Corney, an engineer living in the United Kingdom who used to work for British Telecom, where he used a mobile phone quite heavily—for perhaps 4 or 5 hours per day—for several years. Not long afterward, in his late 30s, he was diagnosed with Alzheimer's disease, and he is now unemployed because his memory is so poor. He believes he developed Alzheimer's disease as a result of his heavy use of the mobile phone manufactured by his employer.[34]

Could this be true? Given that it is a matter of record that exposure to microwave radiation can produce short-term memory loss that lasts for many weeks or months, it is quite reasonable to assume that a person who experienced a very lengthy exposure of this type might suffer permanent or very long-term loss of his short-term memory. In the absence of other experiences that might account for this loss of short-term memory, such a person might very well receive a diagnosis of Alzheimer's disease from a physician he consulted.

Implications for the Establisment of Health Protection Standards

The underlying assumption, in the development of every standard intended to protect human health from exposure to non-ionizing electromagnetic radiation, has been that the higher the radiation intensity, the greater the hazard, as illustrated in Figure 2. This is certainly true for radiation intensities in the range from 10 to 100 mW/cm^2, where the only hazard known is a thermal hazard. The equation for W, the rate of heat generation within a volume of matter, which is derived in this paper, shows that this should be true, and the experimental studies conducted prior to 1966, when ANSI C95 was first issued, confirm it.

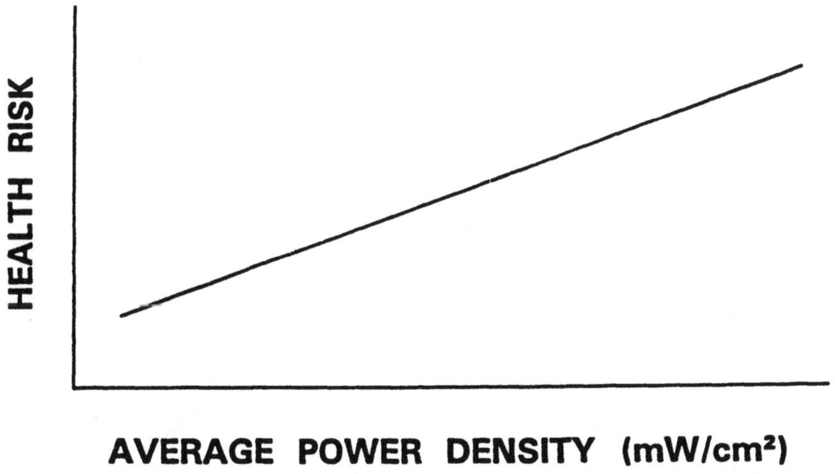

Figure 2. Illustration of a monotonic increasing function (defined as a function, the slope of which is everywhere positive) of average power density. This is the simplest such function, a straight line; it illustrates the relationship between the average power density of microwave radiation and the rate of heat generation in a fixed volume of matter irradiated by this radiation when the radiation source is far away and the matter absorbs virtually all the radiation incident upon it. All existing, recommended, or proposed exposure standards applicable to radio-frequency radiation today are based on the assumption that this is the correct relationship between health hazard and average radiation power density. Notice that this relationship implies that an upper limit imposed on the power density will protect against all health risks.

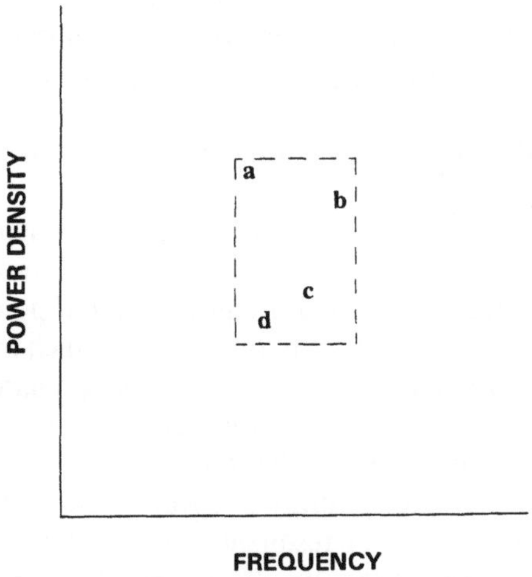

Figure 3. Illustration of a "power-frequency window" indicated by the dashed line: a region of values of power density and a region of values of frequency within which certain types of biological or health effects resulting from non-ionizing radiation exposure seem to be concentrated. The points "a," "b," "c," and "d" represent the values of frequency and power density at which an experiment was done which found abnormal or unusual effects on health; they tend to be clustered together on a plot such as this, giving rise to the idea that there is a "window" of values of frequency and power density within which they occur. A plot of this kind, based on a worldwide search of the literature, was present in the draft criteria document for exposure to microwave radiation which was returned to NIOSH in 1981 by the contractor who held the contract for developing this document at the time this contract was canceled.

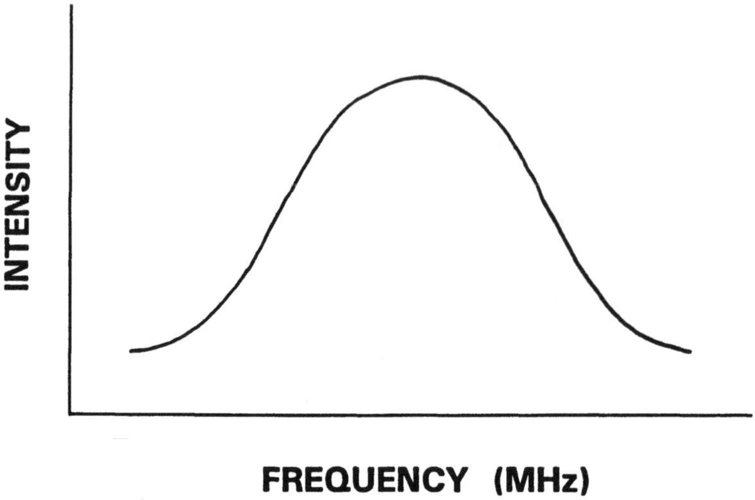

Figure 4. Illustration of a true resonance peak. The strength or intensity of some parameter is displaying resonance when it behaves in this manner as a function of frequency.

But there is abundant evidence from a variety of sources that nonthermal health hazards are largely confined to very low radiation intensities, well below the 10 mW/cm^2 value that electrical engineers have regarded as a safe limit for over 30 years. In the former Soviet Union, and in countries of eastern Europe, values in the vicinity of 1-10 $\mu W/cm^2$ were the official limits for quite a few years in the late 1950s and subsequently.

There have been reports of "power-frequency windows" in the scientific literature for quite some time: reports of abnormal or adverse bioeffects occurring only within specific ranges of values of frequency and of power density, the power density values usually being very low. An illustration of such a "window" is given in Figure 3. Such plots were made by the contractor that prepared the draft document for the US National Institute for Occupational Safety and Health (NIOSH) criteria document for exposure to microwave radiation.

Bioeffects concentrated within a specific frequency range have been encountered before, in the form of resonance phenomena. A resonance curve is illustrated in Figure 4; it shows that an effect is greatly increased within a certain range of frequencies, being smaller on either side of a central frequency. (A simple amplitude-modulated radio circuit exploits the resonance phenomenon to tune the radio receiver to a particular station, so that the signal received at one frequency is amplified more strongly than the signals received at other frequencies; the tuning knob of the radio is connected to a variable capacitor, so that the frequency which is most strongly amplified can be adjusted to "tune in" whatever station is desired.)

Electrical engineers are also aware of this phenomenon. They have taken it

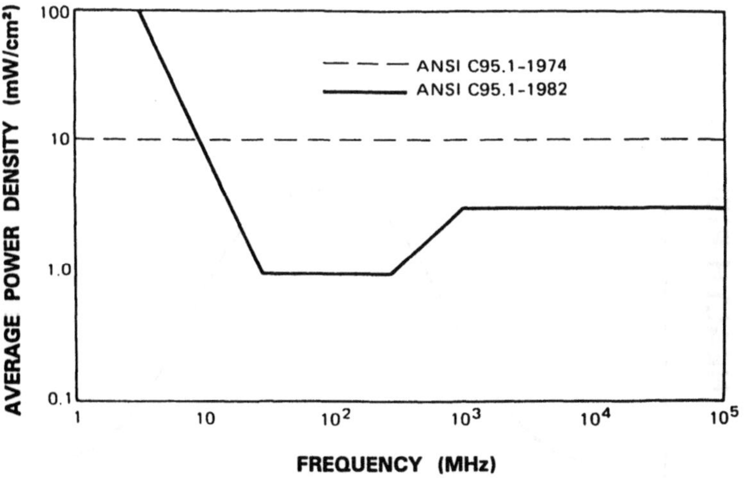

Figure 5. Comparison of frequency-independent ANSI C95-1974 with its 1982 modification designed to compensate for resonance effects, which occur when the wavelength of the radiation is close to the body dimensions (length, height) of an irradiated animal or person. The upper limit on the power density was reduced at those frequencies where resonance occurs in order to compensate for the stronger heating effect that occurs at those frequencies; this is responsible for the "dip" in the curve between 10 and 1000 MHz.

into account in establishing ANSI C95 by lowering the recommended power density over the frequency range where whole-body resonance occurs, which increases the heating inside animals. Figure 5 illustrates this.

But the familiar resonance phenomenon cannot account for the concentration of bioeffects within a particular range of values of the power density. It can only account for a concentration within a particular range of frequencies. Some other phenomenon must be responsible for concentrating nonthermal health effects within a particular range of values of the power density—or within a certain range of values of electric and magnetic field strength, which is more likely to be true.

Whatever the reason for it, nonthermal health effects do seem to be present at very low-power densities and largely absent at the higher power densities near and above 10 mW/cm². This means that over the range from 10 mW/cm² down to some very low value—perhaps near 1 μW/cm², though no one really knows at this time—the nonthermal health hazard increases as the power density is reduced.

Of course, the nonthermal health hazard must drop to zero when the radiation intensity drops to zero. So a risk curve for nonthermal health hazards must have the shape of a peak such as in Figure 6; this is similar to that of a resonance curve (except that it cannot be a true resonance because the horizontal axis represents power density, not frequency).

Is there any experimental evidence to support this concept? One set of experimental data has been published which shows just such peaks; it comes from a study of blood-brain permeability published by Oscar and Hawkins.[35] The graph that was published as Figure 4 in the Oscar and Hawkins paper is reproduced here in Figure 7.

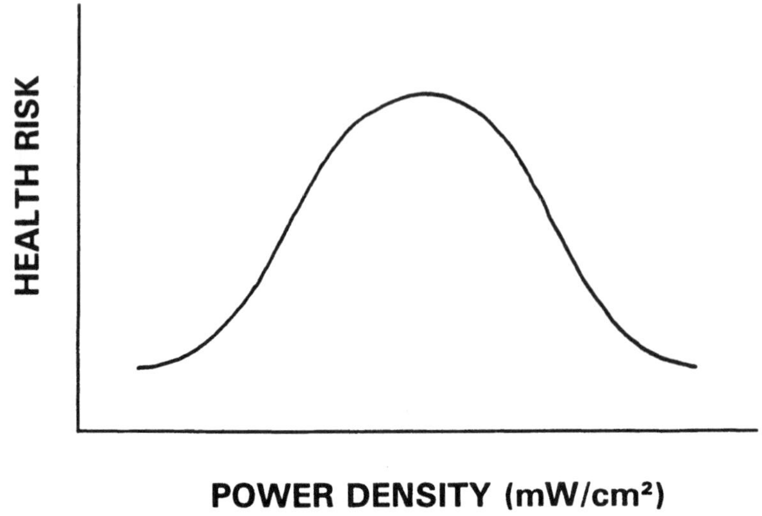

Figure 6. Illustration of a resonance-like peak, showing how nonthermal health effects seem to be concentrated within a certain range of values of the microwave power density. Stochastic resonance is a phenomenon that might be responsible for such an effect.

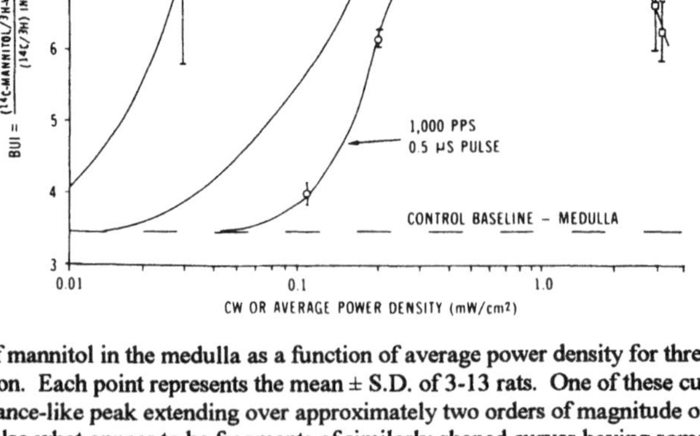

Figure 7. Uptake of mannitol in the medulla as a function of average power density for three types of microwave irradiation. Each point represents the mean ± S.D. of 3-13 rats. One of these curves clearly has the shape of a resonance-like peak extending over approximately two orders of magnitude of the power density. There are also what appear to be fragments of similarly-shaped curves having somewhat different locations and heights. [Reprinted from Oscar KJ & Hawkins TD. Microwave alteration of the blood-brain barrier system of rats. Brain Research 1997; 126: 281-293. Copyright 1977, with permission from Elsevier Science.]

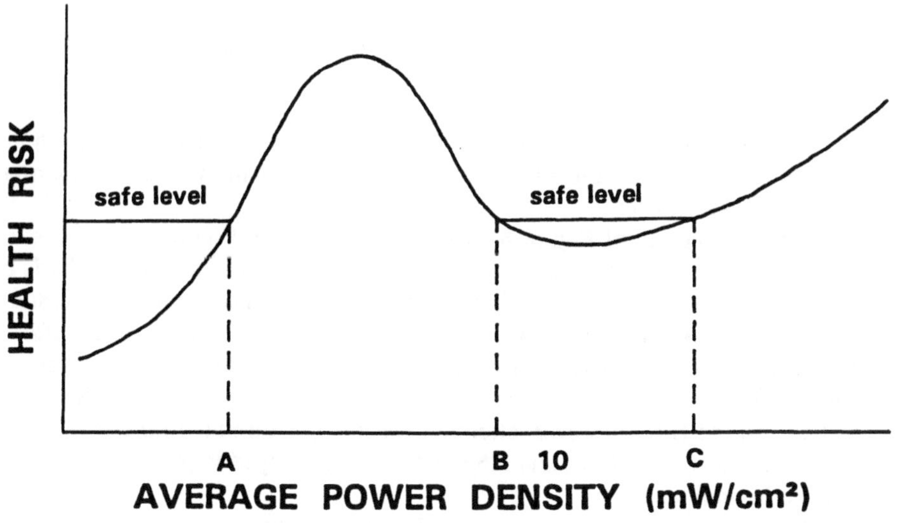

Figure 8. An illustration of the combined thermal and nonthermal heath risk from exposure to microwave radiation, as a function of power density. The peaked portion of this curve represents nonthermal health risks, while the rising portion of this curve above 10 mW/cm^2 represents thermal health risks. Notice that thermal health risks are encountered at power densities above 10 mW/cm^2, while nonthermal health risks are concentrated at power densities below this value. The horizontal line signifies a level of health risk considered acceptable (designated "safe level" on this graph); this line is broken into two segments. One segment covers the range of power density values between zero and the value labeled "A", while the other segment extends from the value of power density labeled "B" to the one labeled "C", where $0 < A < B < 10$ mW/cm^2 $< C$. These two line segments serve to define "safe regions" of exposure to microwave power density. A challenge not yet met is to associate numerical values with the points labeled "A", "B," and "C", assuming that this can indeed be done. Notice that this curve implies that two different upper limits and one lower limit on power density might be established, after values for "A", "B," and "C" have been determined.

When the curve of Figure 6, representing the nonthermal health risk, is added to that of Figure 2, the monotonic increasing curve representing thermal hazard, the result is a curve such as that shown in Figure 8, which illustrates that there are two separate regions of hazard: a thermal hazard at relatively high power densities (certainly above 100 mW/cm^2) and a nonthermal health hazard within a poorly defined intensity range centered at much lower levels (in the mW/cm^2 region). The result is that safe levels of exposure, where both thermal and nonthermal health risks are low, lie at extremely low values of the power density, below both kinds of hazard (and probably well below 1 μW/cm^2), though it appears that there may also be an "island of safety" in the region around 10 mW/cm^2 which lies below the thermal hazard, but above the region of elevated nonthermal risk. Figure 8 illustrates this concept, the horizontal line showing where postulated regions of safety lie.

The curve of Figure 8 has a region in which the nonthermal health risk increases as the power density decreases, while Figure 2, the curve showing a monotonic increase of health risk with increasing power density, has no such region. The region in which a decreasing level of power density is associated with an increasing health risk lies below 10 mW/cm², possibly ranging from 1 mW/cm² down to 1 μW/cm².

Until now, it has been assumed by both industry and government officials that Figure 2 gives a complete description of all health risks, implying that these risks will always be reduced whenever the radiation power density is reduced. In this paper I have presented evidence that it is actually Figure 8 that provides a more nearly correct description. This implies that reducing the power density from mobile phone or other microwave transmitters by a factor of 10 may actually worsen nonthermal health risks such as cancer.

The complexity of Figure 8, when compared with Figure 2, provides an explanation for the wide range of values that have been used as health standards in different countries at different times, illustrated in Figure 1. The countries of eastern Europe were intentionally trying to protect against the subtle nonthermal health effects that occur when people are exposed to low-intensity microwave radiation for extended periods of time, in addition to the thermal effects that occur at higher microwave intensities. In the U.S., however, attention was being paid only to thermal health effects in the 1950s and 1960s, and ANSI C95 was developed to protect against the recognized thermal hazard.

In 1979 Leonard Solon, a health physicist working in the New York City Health Department, tried to provide an increased degree of health protection to the residents of that city by proposing a value substantially lower than the ANSI as an environmental level for the city, on the assumption that Figure 2 was valid. Had his proposal been adopted, however, it might not have yielded the additional degree of health protection he expected, given that the curve of Figure 8 is a better representation of health risk as a function of microwave power density.

It is also unclear whether the very low levels adopted by eastern European countries in the late 1950s and the 1960s would have protected human health, had there been full compliance with them. The health authorities in those countries may also have assumed that they were dealing with a situation such as Figure 2 describes, since they imposed only an upper limit on the radiation intensity.

So far as standards are concerned, all that can be said with confidence is that ANSI C95 and other similar standards based on it, such as those developed by the National Council on Radiation Protection (NCRP), the International Radiation Protection Association (IRPA), the International Commission on Non-Ionizing Radiation Protection (ICNIRP), the American Conference of Governmental Industrial Hygienists (ACGIH), the European Community (EC), and the World Health Organization (WHO), provide no protection at all against nonthermal hazards to health, though they do largely (but perhaps not completely) protect against thermal hazards to health.

Summary of Main Points

1) Nonthermal health effects are real. For a long time their existence was controversial, but they have gradually come to be accepted as real, despite the fact that they are widely regarded at present as being of unknown origin.

Not only are nonthermal health effects real, but they can be explained phenomenologically in terms of the known laws of physics. Precise mechanisms in specific instances remain to be elucidated, but in phenomenological terms, the cause of nonthermal health effects has long been implicitly known—meaning that the theory needed to explain such effects was published and taught many decades ago. The best reference for this purpose is the book[15] first published in 1941 by Julius Adams Stratton, who taught physics at the Massachusetts Institute of Technology, which was regarded in the mid-twentieth century as the premier engineering school in the entire U.S. He laid the foundation for theoretical treatment of nonthermal effects of electromagnetic fields in matter (including in biological systems, though he never explicitly considered them) more carefully than anyone else has done since.

2) Just as thermal health effects arise from the energy-absorbing properties of matter, nonthermal health effects arise from the momentum-absorbing properties of matter. Both these topics are dealt with at some length, both here and in Stratton's book.[15]

3) The exposure metric required to characterize and control nonthermal health risks is different from the exposure metric required to characterize and control thermal health risks. The latter can be derived from an energy balance equation: Poynting's theorem. The former must be derived from a momentum balance equation. Equations involving metrics applicable to local conditions inside a material medium are presented in this paper.

4) There is a range of values of power density lying below 10 mW/cm² within which the risk of nonthermal health effect increases as the power density decreases, illustrated in Figure 8. This must be true if there is a "power-frequency window" at very low intensities within which there is an increased risk of harmful nonthermal health effects. Such "windows" have been reported.

5) There is a small set of apparently safe values of power density—a small "island of safety" within the continuum of values of power density—the limits of which are not well defined at this time, although it does include the value 10 mW/cm². An important task is to define the dimensions of this "island of safety." Its upper limit is determined by the thermal health hazard, while its lower limit is determined by the nonthermal health hazard, as illustrated in Figure 8.

6) Existing standards for human exposure to microwave and radio-frequency radiation, ANSI C95 and all other guidelines or standards developed with reference to it, are inadequate to protect health completely. ANSI C95 was developed to protect against the thermal health hazard of far-field exposure to a source of microwave or radio-frequency radiation. It imposes a limit on the power density—the magnitude of the Poynting vector—which is a metric that, so far as is currently known, can validly be used to protect only against the thermal hazard of a radiation source in the far field of that source.

There is no evidence at this time that the power density, the exposure metric used in all standards and recommendations at this time, validly applies to nonthermal health hazards. The ANSI committee has asserted that ANSI C95 protects against nonthermal hazards to health as well as thermal hazards, but this assertion is based on the assumption that the same metric is valid for protection against both thermal and nonthermal hazards to health—an assumption which the available theoretical evidence does not support, and which is also unsupported by experience.

7) Electrical engineers are not qualified to establish health standards. Properly, all health standards should be established by qualified health professionals. Exposure standards for the purpose of preventing the development of disease or illness should be established by qualified disease prevention professionals. A knowledge of the principles of bioelectromagnetic hygiene should certainly be required on the part of anyone considered to be a qualified disease prevention professional with respect to non-ionizing radiation.

Electrical engineers today are not taught at the undergraduate level about the momentum-related interactions of electromagnetic fields with matter. Mention of the momentum associated with the electromagnetic field is made in some of the physics textbooks[7] used in introductory college and university courses on the theory of the electromagnetic field, though advanced texts[6] must be consulted to learn very much about the absorption of momentum by matter.

8) At this time, no standard has been proposed that would protect credibly against a cancer hazard, which is probably the most serious nonthermal health hazard which exposure to microwave radiation seems to pose. Undoubtedly, it is possible to set a standard so low that it will protect against both nonthermal and thermal health hazards, but what would be a suitable value for such a standard is not known at this time.

In New Zealand Dr. Neil Cherry, who is in the Department of Natural Resources Engineering of Lincoln University, has proposed a very low value—a few microwatts per square centimeter—but there is no evidence that his proposed value will protect against nonthermal health risks. Dr. Cherry has assumed that Figure 2 gives the correct relationship between health risk and radiation power density, an assumption which seems to be false. I fear that Dr. Cherry's recommendation, if implemented, would bring about the exact opposite of what he intends: Instead of greatly reducing the risk of cancer and other nonthermal health effects, these risks would be substantially increased. A standard likely to accomplish what Dr. Cherry intends would probably have to be set from three to six orders of magnitude below the value that he has recommended.

While the safest standard would keep radiation intensity always below the value represented by "A" in Figure 8, it is possible, at least in principle, to reduce the nonthermal health hazard by maintaining intensities at the higher level represented by the values between "B" and "C" in Figure 8, because whatever nonthermal hazards to health exist seem to be encountered almost exclusively at values of the power density below 10 mW/cm^2.

There is no evidence of any nonthermal health hazard to mammals occurring as a result of exposure to microwave or radio-frequency radiation at power densities between 10 and 100 mW/cm^2. There is a possibility of a thermal hazard to health in this range, however, which is why the upper limit of 10 mW/cm^2 was imposed (at a time when the only known hazard to health was a thermal hazard, and it was considered prudent to set an upper limit somewhat lower than might actually be necessary.) It seems likely that the 10 mW/cm^2 limit could be raised somewhat without creating any serious health risk.

The lack of a standard capable of protecting against nonthermal health hazards makes it more difficult to protect human health than would be the case if there were a standard, but it does not make the task impossible. The way to protect people when there is no standard is to keep them far away from the agent posing the hazard. This means that the way to protect people from microwave radiation health hazards at this time is to keep people far away from the microwave radiation—or to keep the microwave radiation far away from where people spend their time.

Existing US voluntary consensus standards do protect human beings from thermal health hazards, so long as they remain in the far field of sources and there are not many sources simultaneously operating. But no standard exists at present anywhere in the world, adopted or proposed, that demonstrably protects human beings against potential nonthermal health effects, such as cancer, that may result from exposure to microwave or radio-frequency radiation.

9) An epidemic of brain cancer readily attributable to cellular telephone use is almost certainly occurring among long-term users of analog cellular telephones. The earliest evidence of such an epidemic consisted of the brain cancer cases that stimulated lawsuits against cellular phone manufacturers and service providers in the U.S. (Similar lawsuits are now beginning to be filed in South Africa and the United Kingdom.)

The brain cancers attributable to the use of analog cellular phones are those malignant tumors occurring in the region of the brain that theoretical evaluation shows to be the region of highest nonthermal health hazard: slightly behind and/or above the ear on the side of the head that the person habitually used the cellular phone.

What is unknown is not whether such an epidemic is occurring, but its magnitude—that is, the number of people affected—as well as the conditions of cellular telephone use that are associated with the development of these cases of brain cancer. Unfortunately, the authorities have been as uninterested as the cellular telecommunications industry in studying these cases of brain cancer allegedly caused by cellular phone radiation. As a consequence of this indifference, the logical initial step toward scientific study of this probable new health hazard—studying the afflicted individuals—has not been taken.

Physicians learn about disease by studying sick people. Health officials would learn about the adverse health effects of exposure to non-ionizing radiation if they were willing to study those individuals who have reported diseases attributable to non-ionizing radiation exposure. In the U.S., neither cellular telephone users alleging brain cancer from their use of these devices nor law enforcement officers claiming cancer caused by their use of traffic radar guns have been studied by any health professional whose purpose was disease prevention and the protection of the public health. The only health professionals who have paid these individuals any attention were the physicians from whom they sought treatment. As a result of the official indifference of the public health authorities, almost nothing is known about the circumstances under which these cancers develop in response to microwave irradiation, despite the deaths of, in all probability, over 200 people in the last two decades in the U.S. from cancer or other diseases associated with exposure to low-intensity microwave radiation.

In addition to medical and hygienic studies of individuals, epidemiological studies are needed. Such studies, properly designed, could confirm the existence of a brain cancer hazard in the population of users of analog mobile telephones, and perhaps could determine whether there is also a brain cancer hazard in the population of users of digital mobile telephones. If any such studies have been carried out, the results have not yet been published.

10) It seems unlikely that a brain cancer hazard will be found in epidemiological studies of users of digital cellular/mobile telephones, because the hazard associated with digital radio-telephone use is distributed across the entire brain, whereas the hazard from analog radio-telephone use is localized in one particular region of the brain. (This does not rule out the possibility of a brain cancer hazard for users of digital mobile telephones; it simply suggests that it may be more difficult to make an association between the use of digital radio-telephones and the development of brain cancer as a consequence, than between the use of analog phones and the development of brain cancer as a consequence.)

There is certainly less likelihood of identifying particular brain cancer cases as being candidates for having been caused by use of a mobile radio-telephone. All types of brain cancer would have to be studied in a population of digital mobile telephone users, whereas only tumors in a particular region of the brain would be relevant in the case of users of analog mobile phones.

Precisely because the hazard from a digital radio-telephone is spread over the entire volume of the brain, those physiological functions such as memory which are not localized in one particular part of the brain would seem to be the most likely to be affected by the radiation from a digital radio-telephone. Loss of short-term memory has been reported in men who experienced accidental over-exposure to aircraft radar at a low intensity level,[29-30] so this is certainly a logical candidate for further study.

Loss of short-term memory in an extremely severe form could be expected to manifest as Alzheimer's disease—which seems actually to have happened to a British engineer in his late 30s who used a mobile telephone very heavily on the job and who a few years later was out of work with a diagnosis of Alzheimer's disease. Therefore early-onset Alzheimer's disease would be another excellent candidate for epidemiological study. Perhaps early-onset Alzheimer's disease (occurring in people below the age of 50) ought to be made a reportable disease, to ensure that statistics are kept on it in a timely manner.

References

1. Solon LR. A local health agency approach to a permissible environmental exposure level for microwave and radiofrequency radiation. Bulletin of the New York Academy of Medicine 1979 Dec; 35(11): 1251-1266.

2. Steneck NH. The microwave debate. Cambridge, MA: MIT Press; 1984. Chapters 3 and 4.

3. Minkowski H. Die Grundgleichungen für die elektromagnetischen Vorgänge in bewegten Körpern. Mathematische Annalen 1910; 68: 472-525.

4. Abraham M. Zur Frage der Symmetrie des elektromagnetischen Spannungstensors. Annalen der Physik 1914; 44: 537-544.

5. Møller C. The theory of relativity. Oxford: University Press; 1957. Pages 202-206.

6. Portis AM. Electromagnetic fields: Sources and media. New York: John Wiley & Sons; 1978. Pages 393-399.

7. Wangsness RK. Electromagnetic fields. 1st ed. New York: John Wiley & Sons; 1979. Pages 422-437.

8. Velizarov S, Raskmark P, & Kwee S. The effects of radiofrequency fields on cell proliferation are non-thermal. Bioelectrochemistry and Bioenergetics 1999 Feb; 48(1): 177-180.

9. Chou CK, Guy AW, Kunz LL, Johnson RB, Crowley JJ, & Krupp JH. Long-term, low-level microwave irradiation of rats. Bioelectromagnetics 1992 June; 13(6): 469-496.

10. Lehmann EL. Testing statistical hypotheses. 2nd ed. New York: John Wiley & Sons; 1986. Page 155.

11. Repacholi M, Basten A, Gebski V, Noonan D, Finnie J, & Harris AW. Lymphomas in Eμ-*Pim1* transgenic mice exposed to pulsed 900 MHz electromagnetic fields. Radiation Research 1997 May; 147(5): 631-640.

12. Fisher PD. Microwave exposure levels encountered by police traffic radar operators. IEEE Transactions on Electromagnetic Compatibility 1993 Feb; 35(1): 36-45.

13. Davis RL & Mostofi FK. Cluster of testicular cancer in police officers exposed to hand-held radar. American Journal of Industrial Medicine 1993 Aug; 24(2): 231-233.

14. Volkers N. Speed kills—But what about radar guns? Journal of the National Cancer Institute 1992 Sept 2; 84(17): 1310-1311.

15. Stratton JA. Electromagnetic theory. New York: McGraw-Hill Book Company, Inc.; 1941. Page 132.

16. Hammett WF. Radio frequency radiation: Issues and standards. New York: McGraw-Hill; 1997. Pages 80-81.

17. Stratton JA. Electromagnetic theory. New York: McGraw-Hill Book Company, Inc.; 1941. Page 100.

18. Abraham M & Becker R. The classical theory of electricity and magnetism. New York: Hafner Publishing Company, Inc.; 1932. Pages 104-108.

19. Braune S, Wrocklage C, Raczek J, Gailus T, & Lücking CH. Resting blood pressure increase during exposure to a radio-frequency electromagnetic field. Lancet 1998 June 20; 351: 1857-1858.

20. Radiation rumor may be probed. Electronics 1971 Aug 2; 44(16): 17.

21. Agencies react to electromagnetic radiation risks. Electronics 1971 Aug 16; 44(17): 35-36.

22. Frey AH. Headaches from cellular telephones: Are they real and what are the implications? Environmental Health Perspectives 1998 March; 106: 101-103.

23. Hocking B. Preliminary report: Symptoms associated with mobile phone use. Occupational Medicine (Oxford) 1998 Sept; 48(6): 357-360.

24. Top scientists give up risky mobile phones because of microwave radiation. Sunday Times (London) 1999 Feb 28.

25. Kraus JD. Antennas. New York: McGraw-Hill Book Company; 1950. Pages 6-7.

26. Silver S. Microwave antenna theory and design. New York: McGraw-Hill Book Company; 1949. Pages 198-199.

27. Lundquist M. Need for categorization of exposure in studies of cellular telephone users. Epidemiology 1999 May; 10(3): 347.

28. Rothman KJ, Chou CK, Morgan R, Balzano Q, Guy AW, Funch DP, Preston-Martin S, Mandel J, Steffens R, & Carlo G. Assessment of cellular telephone and other radio frequency exposures for epidemiological research. Epidemiology 1996 May; 7(3): 291-298.

29. Forman SA, Holmes CK, McManamon TV, & Wedding WR. Psychological symptoms and intermittent hypertension following acute microwave exposure. Journal of Occupational Medicine 1982 Nov; 24(11): 932-934.

30. Castillo M & Quencer RM. Sublethal exposure to microwave radar. Journal of the American Medical Association 1988 Jan 15; 259: 355.

31. Preece AW, Iwi G, Davies-Smith A, Wesnes K, Butler S, Lim E, & Varey A. Effect of a 915-MHz simulated mobile phone signal on cognitive function in man. International Journal of Radiation Biology 1999 April; 75(4): 447-456.

32. Leake J. Mobile phones 'slow' the brain in new tests. Sunday Times (London)1998 Sept 20: 7.

33. Norton C & Woods R. Are we being told the truth about mobile phones? Sunday Times (London) 1998 Dec 20: 10.

34. *The mobile mystery*. Panorama (television broadcast, BBC1, UK). 1999 May 24.

35. Oscar KJ & Hawkins TD. Microwave alteration of the blood-brain barrier system of rats. Brain Research 1977 May 6; 126: 281-293.

Appendix A

The Maxwell stress tensor is defined only in linear isotropic homogeneous media, which includes a vacuum. James Clerk Maxwell discovered and studied it.

The Maxwell stress tensor shows that the presence of electric and magnetic fields produces stresses in space. The scientists of the nineteenth century were familiar with stresses from elasticity theory, which deals with stresses, strains, and small deformations in solid bodies that behave elastically when subjected to forces. To their minds, it was essential for the existence of stresses that a material medium be present to sustain these stresses. The Maxwell stress tensor showed that there are stresses everywhere that electric or magnetic (or electromagnetic) fields are present, therefore nineteenth-century scientists reasoned that there must also be an elastic medium everywhere capable of sustaining a stress, even in a vacuum. This was a major reason why they postulated the existence of an "æther," the sole function of which was to provide a medium through which the electromagnetic field could propagate, when it was not traveling through a condensed material medium.

In the early twentieth century, when experiments disproved the existence of an æther, the concept was abandoned. Physicists today have no difficulty with the concept of stresses existing in a vacuum in the absence of any medium to sustain them, probably because of long familiarity with the theory of relativity.

The Maxwell stress tensor **T** can be most conveniently expressed in terms of the tensor product of pairs of vectors:

$$\mathbf{T} = \mathbf{DE} + \mathbf{BH} - u\mathbf{I}$$

where **I** is the identity tensor, and a tensor product is indicated by writing a pair of vectors side by side without any symbol between them. When the vectors are represented in Cartesian coordinates, it is easy to form a tensor product—each vector is represented as the sum of its vector components, and the two vectors are then algebraically multiplied, component by component, in the same way that two polynomials are multiplied—and it is also easy to take the divergence of the resulting tensor, as shown below.

The Maxwell stress tensor may be written as follows in Cartesian coordinates, where u is the electromagnetic energy density [$u = \frac{1}{2}(\mathbf{D}\cdot\mathbf{E} + \mathbf{B}\cdot\mathbf{H})$] and $\hat{\mathbf{x}}$, $\hat{\mathbf{y}}$ and $\hat{\mathbf{z}}$ are unit vectors of a right-handed Cartesian coordinate system in three-dimensional space:

$$\mathbf{T} = \begin{array}{llllll} + D_xE_x + B_xH_x - u & \hat{\mathbf{x}}\hat{\mathbf{x}} & +D_xE_y + B_xH_y & \hat{\mathbf{x}}\hat{\mathbf{y}} & +D_xE_z + B_xH_z & \hat{\mathbf{x}}\hat{\mathbf{z}} \\ + D_yE_x + B_{y\ x} & \hat{\mathbf{y}}\hat{\mathbf{x}} & +D_yE_y + B_yH_y - u & \hat{\mathbf{y}}\hat{\mathbf{y}} & +D_yE_z + B_yH_z & \hat{\mathbf{y}}\hat{\mathbf{z}} \\ + D_zE_x + B_zH_x & \hat{\mathbf{z}}\hat{\mathbf{x}} & +D_zE_y + B_zH_y & \hat{\mathbf{z}}\hat{\mathbf{y}} & +D_zE_z + B_zH_z - u & \hat{\mathbf{z}}\hat{\mathbf{z}} \end{array}$$

and the divergence of **T** may be taken by thinking of each of the three columns above as a vector in association with a single unit vector. The divergence of **T** is that vector whose x-component is the divergence of the leftmost "column vector" (which is associated with the unit vector $\hat{x}$), whose y-component is the divergence of the central "column vector" (which is associated with the unit vector $\hat{y}$) and whose z-component is the divergence of the rightmost "column vector" (which is associated with the unit vector $\hat{z}$). For example, $[\text{div } \mathbf{T}]_x$—the x-component of the divergence of **T**—is given by the equation

$$[\text{div } \mathbf{T}]_x = \text{div } [(D_xE_x + B_xH_x - u)\, \hat{x} + (D_yE_x + B_yH_x)\, \hat{y} + (D_zE_x + B_zH_x)\, \hat{z}].$$

Appendix B

ANSI C95 came into existence in 1966 as a thermal standard, that is, it was designed to prevent too much heat from being generated within the tissues of a human being or other animal. This is not disputed.

However, claims are now being made that it is no longer a thermal standard. One electrical engineer, William Hammett, has written[16] that ANSI C95-1992 "is distinctly not a 'thermal standard,' a characterization often unfairly attributed to it by otherwise knowledgeable people" indicating that the 'otherwise knowledgeable people' refers to staff of the Environmental Protection Agency who submitted comments to FCC Docket 93-62. Mr. Hammett then goes on to emphasize that the electrical engineers who were concerned with the content of ANSI C95 considered effects in 14 different biological categories. He writes, "The standard was not designed to limit a person's temperature increase to some amount that had been declared unsafe; it is based upon prevention of all reported and substantiated effects, whether they were directly temperature-related or whether the primary effect was nonthermal."[16]

Interestingly, the IEEE has never considered cancer to be a substantiated bioeffect, despite the fact that every controlled lifetime study of laboratory animals exposed to microwave radiation ever conducted has produced highly statistically significant results indicating that an increase in cancer incidence is associated with such microwave exposure. Cancer is a nonthermal health effect[8] which has certainly occurred in populations exposed in compliance with ANSI C95.

Despite the claim that ANSI C95 is no longer a thermal standard, it nevertheless does remain a thermal standard for the following reason: it imposes an upper limit on the magnitude of the Poynting vector: the power density of the radiation. There exists a scientific basis to support the claim that an upper limit imposed upon the magnitude of the power density will prevent a thermal hazard to health. That scientific basis has been presented in this paper.

In contrast, there does not exist any scientific basis to support the assertion that an upper limit imposed upon the power density, at the level that ANSI C95 imposes it, will protect against any nonthermal health hazard. This very issue has been addressed in this paper, and it has been shown, theoretically, that nonthermal health effects are not a function of radiation power density. Furthermore, there is evidence that people exposed to microwave radiation at power density levels in compliance with ANSI C95 have developed cancer that is attributable to their microwave radiation exposure: the law enforcement officers who used the early model traffic radar guns in their work.

The assertion of Hammett and others who insist that ANSI C95 cannot be characterized any longer as a thermal standard is clearly a false assertion. The scientific truth is that ANSI C95 can and does protect only against thermal hazards to health—and it can be relied upon to do so only in the far field of a radiation source, where the plane wave approximation is valid.

II

CERTIFICATION

5

DEVELOPING A REGULATORY COMPLIANCE ARRANGEMENT FOR ELECTROMAGNETIC RADIATION EXPOSURE FROM RADIO TRANSMITTERS BASED ON HEALTH AND SAFETY CONSIDERATIONS IN AUSTRALIA

R.J. Johnston

Abstract

This paper reports on the Australian Communications Authority (ACA) project to introduce compliance arrangements for Electromagnetic Radiation (EMR) exposure from radio transmitters. In June 1997 the legislative basis for technical standards, under the Australian Radiocommunications Act of 1992, was amended to include the power to limit the emission of EMR from radio transmitters based on health and safety considerations. In November 1997 the ACA consulted with communication industry stakeholders and community groups on the principal elements of the compliance arrangements: the public exposure limits in the Australian exposure standard and the compliance demonstration arrangements based on self-declaration. In May 1998 the ACA's Radio Communications Council (RCC) Working Group on electromagnetic compatibility and electromagnetic radiation adopted a management plan and timetable to develop and introduce an EMR compliance regime within 2 years. The management approach for developing the compliance arrangements is open and consultative and includes community and industry stakeholders.

Radio transmitters are divided into two groups for compliance purposes: those whose performance could reasonably be controlled by manufacturers, for example, devices with integral antennas such as mobile phones, and secondly, those whose performance is a function of the installation and operation of the transmitter, for example, mobile base stations. For the first group manufacturers, importers, and agents will have responsibility for the compliance arrangements; they will include labeling with a compliance mark, making a self-declaration statement, maintaining a compliance folder, a regulatory standard (including EMR requirements and test methodology), accreditation for test laboratories, evidence of compliance (test reports), and random auditing. For the second group licensees, through condition of licence, must comply with the requirements in a regulatory standard and the compliance arrangements. These include a self-declaration statement, use of approved laboratories for conducting measurements, evidence of compliance (test or assessment reports), and random auditing. For each group of radio transmitters the extent to which the elements of the compliance arrangements are applied is limited by further categorisation of the radio transmitters based on the level of emission.

On 1 February 1999 the ACA introduced mandatory EMR requirements for mobile phones and their base stations using self-declaratory compliance principles.

Regulatory Environment

Legislation

In 1997 Australia's telecommunication and radiocommunications regulatory scene changed. The Australian government enacted legislation to

1. Replace AUSTEL and the Spectrum Management Agency (SMA) with a single regulatory body, the Australian Communications Authority (ACA);
2. Introduce a new Telecommunications Act of 1997, containing a self-regulatory approach for telecommunications industry regulation; and
3. Change the Radiocommunications Act of 1992 to provide additional power to the ACA to mandate standards that would limit human exposure from radiofrequency (RF) electromagnetic radiation (EMR).

Before the Telecommunications Act of 1997 was introduced manufacturers and importers were required to obtain an equipment permit for devices before they could be sold or connected to a telecommunication network. Permits were obtained by submitting evidence of compliance with relevant regulatory standards to AUSTEL for approval. Under the new self-regulatory industry model manufacturers and importers are obliged to self-declare compliance for equipment and maintain evidence supporting that declaration. To ensure these obligations are met the ACA undertakes a program of random audits.

The changes enacted in the ACA, Telecommunications legislation, and Radiocommunications legislation determined the possible scope for all transmitters, and styles self-regulatory for the EMR regulatory compliance arrangements that have been introduced and which are under development by the ACA. The changes enacted address some of the public concern on the issue of EMR.

In Australia, like the United States and Europe, research is underway into the possible health effects from EMR. The Australian Government's Electromagnetic Energy (EME) Health Issues Committee has set up a $4 million 4.5-year research program into the possible health effects from EMR exposure, this work is managed by the National Health and Medical Research Council (NHMRC). The EME Health Issues Committee also provides public information on EMR issues. In summary, the Australian approach to EMR exposure issues has combined legislating standards and compliance arrangements, providing public information, and undertaking research into health effects.

Information Paper (Step 1)

Details

The ACA's first step towards developing the EMR compliance arrangements was to consult with industry, users, the general public, and different levels of government on a suitable approach. In October 1998 the ACA issued an information paper[1] and sought comment on the document for 2 months. The paper was widely advertised and posted on the ACA's website, and public seminars were held in Sydney and Melbourne to provide a further opportunity to comment.

Later adopted, the paper proposed that the current Australian EMR standard[2] would form the basis of an ACA mandatory standard[3] that set maximum EMR exposure limits for the general public. The mechanisms for implementing compliance with the standard[3] would follow a dual approach that applied the mandatory standard to radiocommunications licences and at the point of supply to the market. The standard[2] is based on the International Radiation Protection Association (IRPA) EMR exposure recommendations circa 1988.

Submissions

The ACA received 43 submissions on the information paper: 8 from industry, 2 from government, 6 from users, and 27 from community groups. The responses to the information paper generally supported the standard[2]-based compliance arrangement proposed. The submissions raised a range of technical and implementation questions that are being addressed in the ongoing development of the compliance arrangements. Public interest groups raised concerns about the possible low-level exposure health effects from EMR. Coincidentally, the joint committee, Standards Australia and New Zealand, responsible for the standard[2] commenced a review of the document.

In March 1998 the ACA, after considering all the submissions, accepted the recommendations in the standard[2] as the basis for developing the EMR compliance arrangements. The ACA's EMR project team then proceeded to develop the detailed compliance arrangements.

Compliance at Point-of-Supply for Devices with Integral Antennas

Manufacturers and importers of devices fitted or intended to be fitted with an integral antenna and intended to be used in close proximity to the body are required to comply with the mandatory standard[3]. These compliance arrangements are based on industry self-declaration of conformity with ACA mandatory standards. The compliance arrangements require manufacturers, importers, and agents to

1. Make a declaration that their product complies with the applicable standards;
2. Create and maintain a compliance folder of documentation supporting the claim; and
3. Label the device.

The compliance label is a protected symbol. Manufacturers, importers, and agents must register with the ACA before using the symbol.

The compliance arrangements, with the mandatory standard[3], are similar to the compliance arrangements for devices specified in regulatory standards for telecommunications, radiocommunications, and electromagnetic compatibility (EMC) arrangements. The harmonization of all the compliance arrangements is currently the subject of a major ACA project.

Two categories of compliance requirements are applied to devices with or intended to be fitted with integral antennas and that are intended to be used in close proximity to the human body:

- **Category A** applies to devices that meet the non-evaluation criteria of the mandatory standard[3]. For these devices the transmitted power does not exceed 20 mW. A declaration of conformity is the only documentation required for this category of devices.

- **Category B** applies to devices that require evaluation according to the mandatory standard[3]. Manufacturers and importers of this category of devices are required to make a declaration of conformity based on test reports.

Compliance at Licensed Transmitter Sites by Users and Operators

Compliance is required of operators and users of all transmitters presently licensed under the Radiocommunications Act of 1992 including those used for telecommunications and broadcasting, both commercial and public. In general, two categories of compliance will apply to licensing conditions:

- **Category 1** applies to transmitters whose output meets the non-evaluation criteria. The non-evaluation criteria will be expressed in terms of transmitter parameters, that is, frequency, height, and power. Currently the specification of these criteria is under development. No documentation is required for this category of transmitter.

- **Category 2** applies to transmitters that require evaluation. Compliance may be verified by two routes:
  1. A self-assessment approach using evaluation documentation provided by the ACA, or
  2. A field measurement survey undertaken by a qualified person.

  For this category of transmitter licensees are required to hold a declaration of conformity and the supporting compliance evidence.

A qualified person is either

1. Accredited (per International Standards Organization [ISO] Guide 25) by the National Association of Testing Authorities (NATA), or a body with a MRA with NATA, to conduct tests required to verify the requirements specified in the regulatory standard[3]; or has
2. Completed a registered electromagnetic radiation evaluation course with the Commonwealth Education Department.

Development Plan (Step 2)

Overview

An ACA EMR project team, which was established in 1997, manages the development of the EMR compliance arrangements. The team is comprised of four members who were drawn from the telecommunications and radiocommunications standards groups within the ACA. This project team is responsible for developing the detailed compliance arrangements.

The EMR project team's major responsibilities include

1. Developing the EMR compliance details;
2. Developing and managing the implementation time plan;
3. Consulting and advising the Radiocommunication Consultative Council (RCC) on EMR planning and implementation matters;
4. Developing education campaigns for the public and industry;
5. Establishing and managing task groups as required;
6. Developing regulatory instruments as required to introduce the framework; and
7. Resolving any compliance demarcation issues with the Health Department and Australian Radiation Protection and Nuclear Safety Agency (ARPANSA).

Management and Consultive Structure

In May 1998 the EMR Project team submitted a project plan to the RCC working group on EMC and EMR, which was subsequently endorsed, that proposed a management approach and committee structure for developing EMR compliance arrangements. The management approach was consultative and allowed participation by stakeholders in the development of EMR compliance arrangements.

Figure 1. Project management diagram.

RCC EMR Planning and Implementation Committee.

The RCC EMR Planning and Implementation Committee was established to advise the ACA on planning and implementation matters associated with the project. The committee is comprised of project stakeholders. The RCC EMR Planning and Implementation Committee's role includes advising the ACA on

1. The EMR compliance arrangements;
2. The implementation time plan;
3. Product labeling;
4. The education campaigns for public and industry;
5. Membership and terms of reference (TOR) for task groups; and
6. Task group reports.

To assist the project team in undertaking its responsibilities, ad hoc task groups are established for a period of time to carry out defined tasks. The membership and TOR for task groups are documented and reviewed by the RCC EMR Planning and Implementation Committee. The agreed TOR for task groups is available on the ACA's website. The following task groups are currently assisting the EMR project team:

1. Category 1 and Category 2 and Compliance Tools Task Group;
2. Mobile Phone and Radio Task Group; and
3. Public Education Campaign Task Group.

Category 1 and Category 2 and Compliance Tools Task Group.

The objective of the Category 1 and Category 2 Task Group is to recommend

1. Criteria to delineate the category 1 and 2 installations. This recommendation will have regard to suitable justification of the likelihood of installations exceeding the human exposure standard, the scope of installations that are likely to fall within category 2, and workable administrative arrangements that enable the ACA to identify from official records individual licences or operators of installations that fall within category 2, and
2. Adoption of compliance tools and/or methodologies (e.g., guidance material, charts, graphs, software, etc.) for use by licensees or operators in evaluating category 1 installations against the human exposure standard.

When defining the categories, consideration will be given to situations where multiple devices are located at a site and/or multiple sites are in close proximity. The ACA is employing specialist consultants to assist this task group to develop the rules for category 1 and category 2 delineation and the assessment tools.

Mobile Phone and Radio Task Group.

In order to evaluate compliance of Category B devices with the RF exposure limits in the mandatory standard, the ACA has nominated a suitable SAR test methodology and documented the rationale for the choice. The Mobile Phone and Radio Task Group was established to assist the ACA with the selection process.

The activity of the task group was required because an international standard specific absorption rate (SAR) test methodology was not be available or likely to be so for several years.

The objectives of the task group were to

1. Review and select from currently available methodologies;
2. Document the selected methodology and a suitable report form; undertake public consultation; and
3. Complete the task by December 1998.

Public Education Campaign Task Group.

This task group concentrated on issues relating to the implementation of the ACA's new regulatory framework.

In this context, the task group on public education about the new ACA EMR regulations has a number of functions:

1. To contribute to publicly acceptable solutions on such issues as labeling/signage and a timetable of phase-in of the regulations. Other task groups will be looking at these issues from an industry perspective;
2. To assist with determining which compliance issues should be the subject of a public education campaign;
3. To advise on the content of materials which will be the backbone of the campaign and where possible assist with the development of these materials;
4. To advise on strategies to relay this information to the target audience; and
5. To evaluate the effectiveness of the campaign.

Time plan

In the first instance, the ACA has applied the non-occupational (general public) exposure limits of the standard[3]. The occupational limits will be considered when the phase-in of the implementation of the non-occupational limits has sufficiently progressed.

Mandatory compliance requirements with the non-occupational limits of the standard[3] are being phased in over a period that is still being determined in consultation with stakeholders. This process will ensure that the appropriate testing infrastructure has been established and is available for all radiocommunications transmitters. The phase-in arrangements are as follows:

- **1 February 1999: Mobile Telephony**
  - Mobile phone handsets (AMPS, GSM 900, and CDMA)
  - Corresponding AMPS, GSM, and CDMA base stations
  - Cordless phone handsets and cradles (CT2/CT3, DECT, and PHS)
- **Late 1999: Other Devices**
  - Satellite phones, GSM 1800
  - Corresponding GSM base stations
  - Spread spectrum cordless phone handsets and cradles
  - Low-interference potential devices
- **2000: Other Devices**
  - The standard[3] will be applied to all category 1 and 2 transmitters. The timing will be dependent on the development and availability of the compliance tools.
  - Application of the standard[3] to other category B devices. This date is subject to the establishment of appropriate industry infrastructure for equipment assessment.

Basal Research (Step 3)

The ACA maintains a database of licensed radiocommunications transmitters called RADCOM. RADCOM is predominately used by frequency assignors for frequency coordination purposes. The database records contain a variety of information: licence number, power, frequency, location, antenna type, polarization, usage category, and so on.

The ACA commissioned a study[4] of the database to assist the Category 1 and Category 2 and Compliance Tools Task Group with its deliberations. The report analysed the 172,447 transmitter records (46,018 sites) in the database on 1 October 1998. The principal focus of the analysis was transmitter power and effective isotropic radiated power (EIRP). Some selected data from the report are presented in Table 1.

EIRP (W) greater than	Number of transmit records	Percentage of total
	172,447	100
1	153, 040	89
10	130,089	75
50	106,725	62
100	66,055	38
1,000	29,431	17
10,000	13,960	8

Table 1. EIRP distribution by transmitter record count.

The data provide insight into the regulatory impact of simple definitions for category 2. For example, if category 2 was defined as 1000 W EIRP, then 29,431 transmitters (17% of the total 172,447 transmitters) would be subject to the category 2 compliance arrangements. If category 2 was defined as sites with more than 20 transmitters, then 943 sites (2.03% of the total 46,018 sites) would be subject to the category 2 compliance arrangements.

The delineation rules for category 1 and category 2 are under consideration. The task group is considering an Australian version of Table 2 in the US Federal Communications Commission's (FCC) Office of Engineering Technology (OET) Bulletin 65 as a likely alternative; this table specifies power (EIRP) and antenna height data for different transmitter types.

EMR Legislation (Step 4)

Labeling

The ACA introduced two new pieces of legislation to give effect to the new compliance arrangements, a Regulatory Standard[3] and a Labeling Notice[5]. In brief the Regulatory Standard[3] specifies

1. The devices (category A and B) and transmitters (category 1 and 2) required to comply;
2. The maximum exposure limits for EMR; and
3. A SAR test methodology.

The Mobile Phone and Radio Task group developed the SAR test methodology in approximately 8 months. A draft document was published and distributed for public comment. All the comments received were considered before the standard was finalised. In brief the Labeling Notice[5] includes

1. The commencement date for the new compliance requirements for new products supplied to the market after 1 February 1999;
2. A phased introduction: a delay (12 months) in the application of the new compliance arrangements to products being supplied to the market before 1 February;
3. Compliance level 1 and 2 definitions including laboratory accreditation (per ISO Guide 25) and/or quality certification (per ISO 9003) requirements;
4. A declaration of conformity pro forma;
5. Requirement to affix a compliance label to the device;
6. Registration arrangements to use the compliance label;
7. Record-keeping arrangements;
8. Records to be kept for variants of a device;
9. Provision of documents for inspection; and
10. Production of a device for inspection.

Licensing conditions

The license conditions for EMR requirements are currently under consideration but should at least include

1. The commencement date;
2. Any qualifications or phase-in arrangements to the commencement date;
3. A definition for compliance levels A and B and assessment requirements associated with category B;
4. Accreditation and training requirements for laboratories and undertaking category B compliance assessment;
5. A declaration of conformity;
6. Record-keeping arrangements;
7. Provision for an agent to act on behalf of a group of licensees operating at one site; and
8. Provision of documents for inspection.

Auditing

The ACA will undertake random auditing to ensure EMR compliance arrangements are observed. The ACA maintains a supplier code database and the RADCOM database in which details of importers, manufacturers, and agents for devices and licensees for transmitters are recorded. For EMR compliance purposes the ACA will select a percentage of the entries in each database and undertake compliance audits. It is envisaged that auditing all the database entries could take 3 to 5 years and would be resource dependent.

Conclusion

The first phase of the ACA's compliance arrangements for limiting EMR exposure was introduced on 1 February 1999. The arrangements require mobile phones and their base stations and cordless phones to comply with requirements not to exceed the maximum exposure levels in a regulatory standard[3].

In Australia the introduction of the new compliance arrangements were well received in the media; in addition, the education campaign conducted by the ACA for mobile manufacturers appeared to be reasonably successful. Only a limited amount of questions from industry and assistance were sought immediately after the introduction of the new compliance arrangements. The ACA plans to conduct audits for mobile phones shortly to ensure the compliance with the new arrangements. Discussion with industry stakeholders has been scheduled to evaluate the effectiveness of the current arrangements and discuss expanding the scope of the current compliance arrangements.

The new compliance arrangements have been in operation for only 4 months. It is therefore difficult to make a comprehensive assessment of the strengths and weaknesses of the new arrangements. To date, however, the response has generally been positive. The ACA will continue to consult with industry and other stakeholders and address the issues that will arise as we gain more experience with the new compliance arrangements in Australia.

References

1. ACA. Information paper: Radio frequency electromagnetic energy: Mandatory human exposure standards and compliance framework.

2. Australian Standard and New Zealand Standard (AS/NZS) 2772.1 (Interim): 1998 Radio frequency fields Part 1: Maximum exposure levels - 3 kHz to 300 GHz.

3. ACA. Radiocommunications (Electromagnetic radiation - human exposure) standard 1999.

4. Spectrum Engineering Australia Pty. Ltd. A presentation of basal information and recommendations for service delineation to assist with the development of an Australian EMR compliance framework.

5. ACA. Radiocommunications (Compliance labeling - cordless and mobile phones) notice 1999.

6

GUIDELINES FOR RADIOFREQUENCY EXPOSURE FROM MOBILE AND PORTABLE DEVICES ADOPTED BY THE US FEDERAL COMMUNICATIONS COMMISSION

R.F. Cleveland, Jr.

Abstract

In 1996 the United States Federal Communications Commission (FCC) adopted new guidelines for evaluating human exposure to radiofrequency (RF) emissions from transmitters authorized and regulated by the FCC. The FCC guidelines are based on exposure limits recommended by the National Council on Radiation Protection and Measurements (NCRP), the Institute for Electrical and Electronics Engineers (IEEE), and the American National Standards Institute (ANSI). The FCC guidelines are applicable to occupational/controlled and general population/uncontrolled exposure categories and for fixed, mobile, and portable transmitters. For the case of mobile and portable devices, localized exposure limits are specified in terms of specific absorption rate (SAR) for devices that are held in close proximity to the user, such as handheld cellular and PCS telephones. The limit that applies to handheld phones is 1.6 W/kg, as averaged over any 1 g of biological tissue. Equipment, such as cellular phones and PCS devices, must demonstrate compliance with this limit in order to be authorized by the FCC for marketing to consumers in the US. This paper will discuss the FCC's policy in this area and the basis for its exposure guidelines.

In the United States, the Federal Communications Commission (FCC) is responsible for licensing and authorizing telecommunications services except for those actually operated by the US Federal government. Many times we are asked the question: Why is the FCC involved with a health and safety issue such as this one? The answer is that even though the FCC is not a health and safety agency, we do have certain responsibilities under an act of Congress, the National Environmental Policy Act of 1969 (NEPA). However, with respect to determining what may constitute environmental significance, we defer to other expert organizations and agencies for specific guidance, for example, on thresholds for safe exposure to radiofrequency (RF) emissions.

Because of the FCC's responsibility under NEPA, in the early 1980s a RF safety program was established in the FCC's Department of Engineering and Technology. At present we have three people in this group who work in this area on a regular basis. Our group includes Mr. Kwok Chan who works at our laboratory and who is responsible for handset certification analysis. Our group focuses on issues related to human exposure from fixed, mobile, and portable transmitting stations and devices and on compliance with FCC exposure guidelines.

Much of the information that is relevant to this issue can be viewed and downloaded from our website (http://www.fcc.gov). You can obtain information there on FCC RF safety activities. You can also download and view decision documents and other publications which I will address. We have also established an information line where the public can call in and get answers to more routine questions and also to request specific FCC publications related to RF exposure and FCC guidelines and policies.

As a bit of background, the FCC adopted new guidelines in 1996 for evaluating human exposure to RF fields. This was the result of a 3-year process initiated in 1993. Pertinent information about this proceeding can be found in ET Docket 93-62; FCC Document Number 96-326; 61 Federal Register 41006 (7 August 1996); and 47 Code of Federal Regulations (CFR) 1.1307(b). The US Telecommunications Act of 1996 had also mandated that the FCC take final action to adopt guidelines by August 1996. The FCC guidelines are based on recommendations from the following expert organizations: the National Council on Radiation Protection and Measurements (NCRP), the Institute of Electrical and Electronics Engineers (IEEE), and the American National Standards Institute (ANSI). The NCRP recommendations were issued in 1986. The IEEE developed, and ANSI adopted, a revised version of its earlier standard. The revised ANSI/IEEE standard is numbered ANSI/IEEE C95.1-1992. Copies of these standards can be obtained from the NCRP or IEEE, respectively.

The FCC's new guidelines relied heavily on advice from federal health and safety agencies including the Environmental Protection Agency (EPA), the Food and Drug Administration (FDA), the National Institute for Occupational Safety and Health (NIOSH), and the Occupational Safety and Health Administration (OSHA). All of these agencies have endorsed the FCC's decision to adopt new guidelines. A supportive letter from the EPA reads that the FCC's approach "addresses our concerns about adequate protection of public health." Furthermore, the FDA wrote that the FCC decision is "appropriate" and "is supported by a consensus of opinion within the federal health agencies."

The FCC adopted two tiers of exposure guidelines. The least restrictive tier is the occupational/controlled tier. The occupational/controlled limits apply in situations in which persons are exposed as a consequence of their employment provided those persons are fully aware of the potential for exposure and can exercise control over their exposure. Limits for occupational/controlled exposure also apply in situations when an individual is transient through a location where occupational/controlled limits apply provided that the individual is made aware of the potential for exposure. The lower tier is for what we term general population/uncontrolled exposure. The general population/uncontrolled limits apply in situations in which the general public may be exposed, or in which persons that are exposed as a consequence of their employment may not be fully aware of the potential for exposure or cannot exercise control over their exposure. These definitions

Frequency range (MHz)	Electric field strength (V/m)	Magnetic field strength (A/m)	Power density (mW/cm²)	Averaging time (minutes)
0.3-3.0	614	1.63	100 *	6
3.0-30	1842/f	4.89/f	900/f² *	6
30-300	61.4	0.163	1.0	6
300-1500	--	--	f/300	6
1500-100,000	--	--	5	6

f=frequency in MHz.
*=Plane wave equivalent power density.
Table 1. Limits for occupational/controlled Maximum Permissible Exposure.

incorporate the terminology used by the NCRP and, also to a large extent, the philosophy of controlled versus uncontrolled found in the ANSI/IEEE standard. These limits for Maximum Permissible Exposure (MPE) apply to fixed transmitters and to what we term mobile devices. The FCC's MPE limits are shown in Tables 1 and 2. The MPE limits are frequency dependent, and there is a five-fold difference, in terms of far-field equivalent power density, between the limits for the general public and the occupational limits. These MPE limits are adopted from the NCRP recommendations for maximum permissible exposure. They are essentially the same as the ANSI/IEEE limits except for the MPE limits for high frequencies above about 1500 MHz. There are also differences at very low frequencies with respect to magnetic field exposure. However, for the frequencies where most exposures occur, the NCRP MPE limits are basically the same as the ANSI/IEEE limits.

With respect to portable and mobile devices, the FCC adopted the 1992 ANSI/IEEE limits for exposure, in terms of specific absorption rate (SAR), which are the same as the NCRP SAR criteria. The allowable localized SAR value for frequencies from 100 kHz to 6 GHz for occupational/controlled is $\leq$ 8 W/kg.

Frequency range (MHz)	Electric field strength (V/m)	Magnetic field strength (A/m)	Power density (mW/cm²)	Averaging time (minutes)
0.3-1.34	614	1.63	100 *	30
1.34-30	824/f	2.19/f	180/f² *	30
30-300	27.5	0.073	0.2	30
300-1500	--	--	f/1500	30
1500-100,000	--	--	1.0	30

f=frequency in MHz.
*=Plane wave equivalent power density.
Table 2. Limits for general population/uncontrolled Maximum Permissible Exposure.

The allowable localized SAR value at the same frequencies for public/uncontrolled is ≤ 1.6 W/kg. The most important value is 1.6 W/kg, averaged over 1 g of tissue, which is the partial-body limit the FCC applies to consumer products intended for public use. This value is the ANSI/IEEE and NCRP partial-body limit for public exposure, and it is the threshold limit we use for evaluating handsets, such as cellular telephones, for safety.

The FCC has established two categories for portable and mobile devices, so that we can differentiate between them based on the way that a device is used. We define mobile devices as devices that are designed to be used in other than fixed locations and will generally be used at least 20 cm from the body of the user. Portable devices are defined as the types of devices, like handsets, that are held against the head and that are designed to be used within 20 cm of the body of a user. The FCC will allow evaluation of mobile devices with respect to MPE limits for field strength and power density. For example, routine evaluation of cell phones such as bag or suitcase type phones would come under this category.

In terms of portable devices that are the handset devices held close to the head, the FCC requires SAR evaluation. The limit for partial-body SAR is, again, 1.6 W/kg, and we require routine evaluation for cell phones, PCS devices, and mobile satellite devices.

The FCC rules for these specifications are found in 47 CFR Section 2.1091 for mobile devices and 47 CFR Section 2.1093 for portable devices. In some cases, a device cannot be easily classified as either a mobile or portable device, for example, modules and desktop devices. In those cases we allow human exposure to be evaluated with either SAR or MPE field and power density limits, and compliance can often be accomplished by operating instructions, installation procedures, or warning labels. There have been a few cases where these means have been used to achieve compliance. Warning labels, for example, would indicate the minimum distance from a device where exposure limits might be exceeded.

The FCC has recognized that there is a need to provide more specific guidance to our applicants and licensees on how to comply with the safety guidelines. Therefore, in 1997, the FCC issued an updated version of our technical bulletin providing guidance on compliance, OET Bulletin 65. We have also issued three supplements to Bulletin 65 which give more specific directed information on compliance for specific types of transmitters and facilities. Supplement A deals with radio and television broadcasting, Supplement B covers amateur radio, and Supplement C, which is the most relevant document for this group, is written for mobile and portable device evaluation. All of these documents can be downloaded from our website.

Even though the FCC has issued this particular supplement dealing with mobile and portable devices, we recognize that there is still a need for more specific guidance on evaluation procedures, especially with respect to handheld devices. Therefore, FCC staff have been active participants in Subcommittee 2 of the IEEE Standards Coordinating Committee 34 (IEEE SCC34) to develop protocols and guidance for evaluating exposure to portable and mobile devices. This committee was formed in 1997 as a result of a workshop sponsored by the FDA and FCC. The chair of the subcommittee is Howard Bassen of the FDA. The purpose of IEEE SCC34 is to develop product performance standards relative to the safe use of electromagnetic energy utilized by specific products in the frequency ranges of 3 kHz-300 GHz. Subcommittee 2 was formed to develop performance standards for evaluating portable wireless devices, such as handheld cellular phones, for compliance with exposure guidelines.

This subcommittee just met for 2 days in Long Beach, and we are making good progress in developing a draft standard. We expect to have a complete draft by the end of this year. This subcommittee is composed of two working groups. The first working group is developing protocols for laboratory test procedures to be used for determining SAR in the human head due to handheld phones. The chair of this working group is also Howard Bassen. The second working group is developing a standard for computational methods to be used in evaluating SAR from handheld phones. The chair of that group is Kwok Chan of the FCC.

The following are several major areas that need to be addressed with regard to handset evaluation. Many of these issues will be dealt with in the upcoming IEEE SCC34 standard. Others are more specific to FCC requirements and are discussed in our technical bulletins and in other FCC publications.

- How much and what type of technical data need to be submitted to demonstrate compliance?
- Specific information on the device under evaluation, including description, classification, antenna configuration, and output power, should be provided.
- Uniform positioning of the handset with respect to the head during evaluation needs to be described.
- Testing protocols need to be specified in terms of variables such as output power frequency, and so on.
- A description of measurement and positioning systems, including information on measurement probes and calibration procedures for the probes and the system itself needs to be provided.
- Techniques used to determine SAR distributions, for example, the peak SAR levels and determination of 1-g averages, should be described.
- The shape, size, composition, and dielectric properties and complexity of materials used for the measurements must be specified. For example, the type of phantom being used, its shape, and its size is critical in evaluating SAR. Our committee is presently spending a lot of time discussing this issue, and we now think we have a good idea of the type of phantom head we will recommend. We are also approaching closure on the exact dielectric properties that will be recommended.
- The validity of numerical computational techniques, such as finite-difference time-domain (FDTD) techniques, in evaluating and demonstrating compliance with guidelines.
- How to specify uncertainty and margin of error.
- Reproducibility of results is critically important.

In general, our goal is to arrive at a point where we have a technically supportable recommended practice that will lead to reproducibility within an acceptable margin of error for handset certification.

That is a summary of the FCC's policies and guidelines in the area of RF safety compliance, as well as current information on the direction in which we are heading with respect to handset certification.

7

FCC EQUIPMENT AUTHORIZATION PROCEDURES FOR RF EXPOSURE COMPLIANCE OF MOBILE AND PORTABLE TRANSMITTERS

K.W. Chan

Introduction

In August 1996, the Unites States Federal Communications Commission (FCC) adopted a *Report and Order* in ET Docket 93-62 amending its rules for evaluating the environmental effects of radiofrequency (RF) electromagnetic fields. Specifically, the FCC adopted new guidelines and procedures for evaluating human exposure to RF emissions from FCC-regulated transmitters and facilities.[1] As a part of this proceeding, new limits were adopted for human exposure to RF emissions from certain mobile and portable devices. Two subsequent FCC Orders refined and clarified the decisions adopted in the original *Report and Order*.[2] A revised version of OET Bulletin 65, "Evaluating Compliance with FCC Guidelines for Human Exposure to Radiofrequency Electromagnetic Fields" was also issued.[3] One of the areas discussed in Bulletin 65 is compliance with the limits adopted by the FCC for safe exposure to RF emissions due to mobile and portable devices such as non-fixed wireless transmitters and handheld cellular telephones. Subsequently, the first edition of Supplement C to OET Bulletin 65 was released in December 1997 to provide applicants with guidance for evaluating RF exposure compliance of mobile and portable transmitting devices.[4] While industry groups and standards organizations are working to develop test protocols for RF exposure evaluation, the FCC has reviewed numerous reports on routine RF exposure evaluation for mobile and portable transmitters. Compliance has been determined according to the test procedures and supporting results on a case-by-case basis reviewed for equipment approval. The specific guidance on test methods and additional clarifications on technical reporting procedures described in Supplement C have provided useful information to parties filing applications for equipment authorization that require routine RF evaluation to demonstrate compliance. However, the information provided in OET Bulletins and supplements issued by the FCC are not intended for establishing mandatory procedures, as other methods and procedures may be acceptable if scientifically sound engineering practices are confirmed.

FCC rules require applicants for equipment authorization of certain portable and mobile devices to provide an affirmative statement of compliance attesting that a device complies with FCC limits for RF exposure. The rules also require that technical information be provided upon request for demonstrating compliance. Because of the lack of any final recommendations issued by accredited standards organizations, the FCC has been requesting technical data on all routine RF exposure evaluations for mobile and portable transmitters. Compliance has been determined according to the test methodology, technical descriptions, and supporting results submitted during equipment approval.

The following sections are brief summaries of the information contained in Edition 97-01 of Supplement C to OET Bulletin 65, "Additional Information for Evaluating Compliance of Mobile and Portable Devices with FCC Limits for Human Exposure to Radiofrequency Emissions." OET Bulletin 65 and Supplement C are available from the FCC website: http://www.fcc.gov/oet/info/documents/bulletins/#65.

FCC Rules for RF Exposure Compliance of Mobile and Portable Transmitting Devices

Mobile and portable transmitting devices that operate in the Cellular Radiotelephone Service, the Personal Communications Services (PCS), the Satellite Communications Services, the General Wireless Communications Service, the Wireless Communications Service, the Maritime Services (ship earth stations only) and the Specialized Mobile Radio Service authorized under Part 22 (Subpart H), Part 24, Part 25, Part 26, Part 27, Part 80, and Part 90 of the FCC rules are subject to routine environmental evaluation for RF exposure prior to equipment authorization or use. Unlicensed PCS, Unlicensed NII, and millimeter wave devices authorized under Part 15 of FCC rules are also subject to routine environmental evaluation for RF exposure prior to equipment authorization or use. All other mobile and portable transmitters are categorically excluded from routine environmental evaluation for RF exposure.[5] However, these transmitters are not exempted for compliance. If a categorically excluded transmitter has the potential for exceeding RF exposure requirements, an evaluation may be requested as provided for by 47 CFR §1.1307(c) and (d).

The FCC rules for evaluating mobile devices for RF exposure compliance are found in 47 CFR §2.1091. For purposes of RF exposure evaluation, a mobile device is defined as a transmitting device designed to be used in other than fixed locations and to be generally used in such a way that a separation distance of at least 20 cm is normally maintained between the transmitter's radiating structures and the body of its user and nearby persons. Transmitters that cannot be qualified under the operating and installation criteria intended for plane-wave exposure described in Table 1 of 47 CFR §1.1307 are generally considered as mobile transmitting devices. Examples of mobile devices, as defined above, would include transmitting devices such as cellular and PCS mobile telephones that use vehicle-mount antennas and certain other transportable transmitters. Mobile transmitters must be evaluated with respect to Maximum Permissible Exposure (MPE) or Specific Absorption Rate (SAR) limits, given in Appendix A and B respectively, to determine compliance. Unless it is desirable to perform SAR, generally it is simpler and more cost-effective to evaluate mobile devices with respect to MPE limits based on field strength or power density to determine compliance.

The FCC rules for evaluating portable devices for RF exposure compliance are contained in 47 CFR §2.1093. For purposes of RF exposure evaluation, a portable device is defined as a transmitting device designed to be used with any part of its radiating structures in direct contact or within 20 cm of the body of its user or other persons under normal operating conditions. This category of devices includes cellular and PCS handheld telephones, wireless transmitters that are carried next to the body of a person, and certain wireless desktop telephones that do not have effective means to allow its users and nearby persons to maintain 20 cm or more from their radiating antenna. Portable transmitters used by consumers are generally required to perform routine environmental evaluation for RF exposure with respect to uncontrolled SAR limits. Compliance is determined according to the highest SAR averaged over any 1 g of tissue, defined as a tissue volume in the shape of a cube.[6]

In general, occupational/controlled exposure limits are applicable to situations where persons are exposed as a consequence of their employment and have been made fully aware of their potential for exposure to exercise control over the exposure. This exposure category is also applicable when exposure is of a transient nature due to incidental passage through a restricted or controlled location where the exposure levels may be higher than the general population/uncontrolled limits. However, the exposed persons must be fully aware of their potential for exposure and can exercise control over the exposure by leaving the area or by some other appropriate means. Exposure awareness can be provided through specific training, as part of an RF safety program, to make persons knowledgeable about their potential for RF exposure in a workplace or similar environment. If it is appropriate, effective warning signs and labels may also be used to establish such awareness by providing prominent information on the risk of potential exposure and to include instructions on useful methods to minimize exposure risks.

The general population/uncontrolled exposure limits are applicable to situations where members of the general public are exposed to conditions not related to their employment, such as nearby persons in the vicinity of a wireless transmitter. This exposure category also includes persons who are exposed as a consequence of their employment and have not been made fully aware of the potential for exposure or cannot exercise control over their exposure. The use of warning labels placed on low-power consumer devices such as cellular telephones and other similar transmitters are not considered sufficient for restricting transmitters to employment-related usage only; compliance must be determined with the general population/uncontrolled exposure limits.

General Evaluation Criteria for RF Exposure Compliance of Mobile and Portable Transmitters

Unless otherwise categorically excluded, mobile devices that operate at 1.5 GHz or below with an effective radiated power (ERP) of 1.5 W or more, and those that operate at frequencies above 1.5 GHz with an ERP of 3 W or more are required to perform routine environmental evaluation for RF exposure prior to equipment authorization or use. Compliance may be evaluated with respect to field strength, power density, or SAR limits, as appropriate. Portable devices must be evaluated with respect to SAR limits using either measurement or computational methods to determine compliance according to the highest SAR averaged over any 1 g of tissue in the shape of a cube.

For routine evaluation, an applicant must determine if a transmitter is a "mobile" or "portable" transmitter according to the operating and exposure conditions of the device. The applicable occupational/controlled or general population/uncontrolled exposure limits for field strength, power density, or SAR should be applied to evaluate compliance for users and bystanders. For certain devices such as CDPD modems, wireless data communication modules, or other similar transmitters that operate as accessories to other electronic products where RF exposure for both mobile and portable operating conditions are applicable, RF exposure compliance should be evaluated with respect to SAR limits for the portable configurations. When exposure is employment related where the operator is in close proximity to a transmitter but nearby persons are normally further away from it, the limits for occupational/controlled exposure may be applied to the operator if the person is aware of his or her exposure conditions and can exercise control to limit the exposure duration and/or conditions to satisfy compliance.[7] For nearby persons who have no knowledge of their exposure conditions, even though further away from the transmitter and exposed to a weaker field, the more restrictive limits for general population/uncontrolled exposure must be used to determine compliance.

SAR limits are not applicable for devices operating above 6 GHz. RF exposure for certain portable devices operating under Part 15 of FCC rules, such as millimeter wave transmitters, must be evaluated with respect to power density limits. Reliable power density measurements can generally be made at 5 cm or more from transmitters operating at shorter wavelengths above 6 GHz. If the normal operation of a device requires less than 5 cm separation from persons, power density computations based on numerical techniques used for SAR evaluations may be used.

For dual mode transmitters that operate in a single frequency band, MPE or SAR may be evaluated for the worst case operating mode according to the peak and average output power levels, modulation characteristics, bandwidth, and performance requirements to determine compliance for all the operating modes. For example, testing in the AMPS mode is sufficient for both AMPS and TDMA modes if the peak output power in TDMA mode does not exceed that transmitted in the AMPS mode. For certain modulations that operate with a rather wider bandwidth, such as direct sequence spread spectrum or CDMA transmitters, the output power and total energy emitted must be considered to determine the worst case operating mode. If the worst case operating condition cannot be confidently determined for a dual mode device, compliance for each individual operating mode should be considered. Since device performance, energy absorption, and MPE/SAR distribution are generally different in separate frequency bands, compliance for devices operating in multiple frequency bands must be evaluated for the operating modes in each frequency band.

The performance of a transmitter may vary within its transmission band according to the design and performance of its antenna and other circuitry. Certain types of antennas, such as helices and patches, could be more sensitive to loading by body tissues and surrounding objects that may cause unintended changes in device performance. To fully evaluate the worst case operating conditions, portable transmitters should generally be tested at the high, middle, and low frequencies of the transmission band. Mobile transmitters that operate in a semi-isolated environment are less likely to experience similar changes; evaluation in the middle of the frequency band is generally sufficient. If a transmitter can operate with a substantially wide range of operating frequencies, more frequencies should be considered if device performance is suspected to vary across the operating frequencies.

With respect to the normal operating positions of a mobile or portable transmitter, if the device has a retractable antenna, compliance should be evaluated with the antenna in its fully extended and retracted positions. Most transmitters generally do not perform well with their antennas partially extended or retracted. If it is suspected that these antenna positions can lead to excessive RF current flow on the outer surface of the device and relatively higher field strength or SAR may be expected, the situations should be analyzed to determine if evaluation is necessary or if appropriate operating instructions and/or warnings could be used to caution users to avoid using these antenna positions.

The applicable procedures for evaluating mobile transmitters with respect to MPE limits are described in OET Bulletin 65 and in the ANSI/IEEE C95.3, "IEEE Recommended Practice for the Measurement of Potentially Hazardous Electromagnetic Fields - RF and Microwave." In general, field strength or power density measurements should be made in all directions surrounding a mobile transmitter, without the influence of nearby persons and objects. The measurements should be spatially averaged over the exposed portions of a person's body corresponding to worst case exposure conditions for the normal operating conditions and configurations of the device. For certain installations, such as vehicle-mounted antennas or those located on a metallic ground plane, exposure should be evaluated with the antenna positioned on an equivalent metallic surface to account for ground reflections that may result in higher localized peak power densities near the antenna.

Portable transmitters should be evaluated with respect to SAR limits using tissue equivalent phantom models. The shape, dimension, and complexity of the tissue model should support worst case exposure conditions for the normal operating conditions and configurations of the test device. Near-field exposure conditions require a phantom to simulate the physical shape and dimensions of the body regions exposed to the transmitter to produce SAR distributions that would closely represent exposure conditions expected during actual use. The RF absorption characteristics of human tissues are frequency dependent. The dielectric parameters of simulated tissue materials used in the SAR evaluation must match the properties of tissues normally exposed to the transmitter, within the operating frequency range of the device, for evaluating worst case exposure for normal use. The tissue equivalent liquid is typically contained in a realistic human-shaped phantom shell, allowing a small diameter electric field probe to move freely within the liquid to assess the highest electric field strength within the tissue regions exposed to the transmitter.

In general, duty factors related to device usage, software programming, or other operational factors that are not inherent in the hardware design and transmission protocol of a transmitter are usually not applicable to source-based time-averaging. The rules only allow time-averaging of typical maximum duty factors for mobile and portable transmitters when controlled exposure limits are applicable; source-based time-averaging must be used when uncontrolled exposure limits are used (see 47 CFR §2.1091(d)(2) and §2.1093(d)(5)).

Evaluating Compliance with Respect to SAR Limits

Localized partial-body exposure requires SAR to be spatially averaged over any 1 g of body tissues, or 10 g for hands, wrists, feet, and ankles, in the shape of a cube for determining compliance with respect to occupational/controlled or general population/uncontrolled exposure limits.[8] Portable transmitters such as wireless handsets and pagers are generally tested in their normal operating positions next to a homogeneous head or body phantom filled with tissue equivalent liquid. The peak SAR locations should be initially searched throughout the tissue regions exposed to the transmitter, at 1-2 cm spatial resolutions, to determine the locations of highest energy absorption. This is subsequently followed by measurements, at 2-3 mm spatial resolutions, at one or more of the peak exposure locations to determine the highest SAR averaged over any 1 g of tissue in the shape of a cube. For homogeneous models, the highest SAR generally occurs closest to the surface of the phantom and is usually not measurable by most electric field probes because the sensors are typically located at 2-5 mm behind the tip of the probe. The highest SAR values at the surface of the phantom must be extrapolated according to measurements taken several millimeters from the surface, at the peak SAR region, to correctly determine the highest 1-g averaged SAR.

Evaluating SAR Using Computational Procedures

The criteria for evaluating mobile and portable transmitters with measurements, such as device positioning, device operating conditions, and tissue modeling requirements, are generally applicable to evaluating RF exposure with computational methods. The majority of SAR evaluations for equipment approval have been based on measurements; only a few have used computational techniques. While much work is needed to clearly define computational test procedures similar to those used in measurements, most of the latest efforts in standards organizations have been limited to developing SAR measurement protocols for testing wireless handsets. A summary of the fundamental concepts for evaluating RF exposure using computational techniques and certain validation procedures used for supporting test results is described in Supplement C to OET Bulletin 65.

Preparing RF Exposure Test Reports for Equipment Approval

MPE and SAR evaluations typically require documentation of certain supporting information to demonstrate a device meets RF exposure limits. This generally includes descriptions of evaluation procedures, device test configurations, measurement results, and explanations of data analysis methods to substantiate compliance. The extent of documentation may vary according to procedures adopted at individual test sites and sometimes depends on the purpose of an evaluation. It is not always necessary to include extensive explanations of all the test procedures used in an evaluation or to provide in-depth technical details in test reports to demonstrate a device meets RF exposure compliance limits. The detail test procedures are usually documented separately according to established measurement protocols adopted by standards organizations or individual test sites. The specific information relating to the evaluation of an individual device being tested should be documented in sufficient detail to allow a test to be repeated independently within its specified error margin.

The MPE or SAR evaluation procedures used by a test facility must be clearly documented, including the protocols used to validate the measurement system and calibrating the supporting equipment to creditable measurement standards and references. A detailed calibration record should be maintained for all the equipment used by a test site. Typically, a brief summary of the calibration records is required to support measurement accuracy. The test protocols and procedures that have been fully documented by creditable organizations should be clearly identified and briefly described in the test report. In general, descriptions of the test device, its operating conditions, phantom and device test configurations, list of equipment, and calibration data should be included in a test report. The results of different test conditions should be tabulated to indicate that worst case exposure has been determined. Plots and illustrations showing the peak exposure location with respect to the setup positions of the tissue phantom and test device should be used to demonstrate the measurement or computational procedures are suitable for evaluating the test device. The test report should also include an uncertainty analysis and a statement of compliance indicating the conditions under which a device has been evaluated for meeting the specified exposure limit and the extent of compliance warranted by the uncertainties determined during the evaluation.

Notes

[1] See *Report and Order*, in ET Docket 93-62, FCC 96-326, 11 FCC Rcd 15123 (1997). The Commission's environmental rules for RF exposure are described in 47 CFR §1.1307(b).

[2] See *First Memorandum Opinion and Order*, ET Docket 93-62, 11 FCC Rcd 15123 (1997). See also *Second Memorandum Opinion and Order and Notice of Proposed Rule Making*, ET

Docket 93-62, released on August 25, 1997.

[3] OET Bulletin 65, Edition 97-01, released August 25, 1997. This document can be downloaded from the FCC's website: http://www.fcc.gov/oet/. Paper copies can be obtained from the FCC's duplicating contractor, International Transcription Service, Inc. at (202) 857-3800.

[4] Subcommittee 2 of Standards Coordinating Committee 34 (SCC-34) sponsored by the Institute of Electrical and Electronics Engineers (IEEE) was formed in February 1997 to develop recommendations with respect to evaluation of portable devices for compliance with SAR limits using experimental or numerical methods. Additional information is available at the IEEE Standards Association website (http://standards.ieee.org) and the SCC-34, SC-2 website: (http://grouper.ieee.org/groups/scc34/sc2).

[5] See 47 CFR § 2.1091 and 47 CFR § 2.1093 for details.

[6] Both the ANSI/IEEE and NCRP exposure criteria are based on a determination that potentially harmful biological effects can occur at a SAR level of 4 W/kg as averaged over the whole body. Appropriate safety factors were then added to arrive at limits for both whole-body exposure (0.4 W/kg for "controlled" or "occupational" exposure and 0.08 W/kg for "uncontrolled" or "general population" exposure) and for partial-body (localized SAR), such as might occur in the head of the user of a handheld cellular telephone.

[7] An example would be certain transmitters used at checkout stands.

[8] See ANSI/IEEE C95.1-1992.

Appendix A

FCC Limits for Maximum Permissible Exposure (MPE)

Frequency range (MHz)	Electric field strength (E) (V/m)	Magnetic field strength (H) (A/m)	Power density (S) (mW/cm^2)
0.3-3.0	614	1.63	100 *
3.0-30	1842/f	4.89/f	$900/f^2$ *
30-300	61.4	0.163	1.0
300-1500	--	--	f/300
1500-100,000	--	--	5

f = frequency in MHz.
* Plane-wave equivalent power density.
Limits for Occupational/Controlled Exposure.

Frequency range (MHz)	Electric field strength (E) (V/m)	Magnetic field strength (H) (A/m)	Power density (S) (mW/cm^2)
0.3-1.34	614	1.63	100 *
1.34-30	824/f	2.19/f	$180/f^2$ *
30-300	2.75	0.073	0.2
300-1500	--	--	f/1500
1500-100,000	--	--	1.0

f = frequency in MHz.
* Plane-wave equivalent power density.
Limits for General Population/Uncontrolled Exposure.

Appendix B

FCC Limits for Specific Absorption Rate (SAR)

Whole-body	Partial-body	Hands, wrists, feet, and ankles
0.4	8.0	20.0

Limits for occupational/controlled exposure (W/kg).

Whole-body	Partial-body	Hands, wrists, feet, and ankles
0.08	1.6	4.0

Limits for general population/uncontrolled exposure (W/kg).

NOTE 1: Whole-body SAR is averaged over the entire body; partial-body SAR is averaged over any 1 g of tissue defined as a tissue volume in the shape of a cube. SAR for hands, wrists, feet, and ankles is averaged over any 10 g of tissue defined as a tissue volume in the shape of a cube.

NOTE 2: At frequencies above 6 GHz, SAR limits are not applicable and MPE limits for power density should be applied.

EDITOR'S NOTE: WTR TIER I

WTR's Tier I toxicology program is designed to be consistent with standard approaches to pre-market product safety evaluation. While there are no requirements for pre-market approval of wireless phones, the WTR program is designed to build the same database that would be required under a standard pre-market approval protocol.

The WTR program has developed new exposure systems for both in vitro and in vivo experiments that allow for assessment of effects in experiments that closely mimic the type of exposures that would be encountered by people using wireless phones. In addition, the WTR exposure systems allow for assessment of dose-response relationships that are critical to judging the public health significance of toxicological experiments.

The following tables list the toxicological experiments relevant to RFR exposure from wireless phones. Table 1 includes those toxicological studies completed within the WTR Tier I program. Tables 2 and 3 list toxicological studies and reviews identified through ongoing literature surveillance.

Overall, current toxicology studies indicate that the energy from wireless phone antennas is not likely to cause heating of tissue or breakage of DNA. However, consistent findings of increased formation of micronuclei in exposed human blood cells suggests genotoxicity not mediated through DNA breakage.

Lab	Biological model	Biological endpoint	Exposure conditions	Results
IN VITRO TOXICOLOGY				
ILS*	bacterial cultures (*salmonella typhimurium/E. coli*)	reverse mutation induction	837 MHz in analog, TDMA, CDMA mode 3 hours at 1, 2.5, 5, 10 W/kg	Not mutagenic
ILS	bacterial cultures (*salmonella typhimurium/E. coli*)	reverse mutation induction	1909.8 MHz in PCS mode 3 hours at 1.0-1.5, 2.5-3.3, 5.0-5.1, 9.6-11.0 W/kg	Not mutagenic
SRI**	bacterial cultures (*salmonella typhimurium/E. coli*)	reverse mutation induction	837 MHz in analog mode 3 hours at 1, 2.5, 5, 10 W/kg	Not mutagenic
ILS	mammalian cell culture (mouse lymphoma cells)	deletions and point mutation induction	837 MHz in analog, TDMA, CDMA mode 3 hours at 1, 2.5, 5, 10 W/kg	Not mutagenic
ILS	mammalian cell culture (mouse lymphoma cells)	deletions and point mutation induction	1909.8 MHz in PCS mode 3 hours at 1.3, 3.3, 5.3, 10.3 W/kg	Not mutagenic
SRI	mammalian cell culture (mouse lymphoma cells)	deletions and point mutation induction	837 MHz in analog mode 3 hours at 1, 2.5, 5, 10 W/kg	Not mutagenic
ILS	human blood lymphocytes	chromosomal aberrations	837 MHz in analog, TDMA, CDMA mode 3 hours at 1, 2.5, 5, 10 W/kg and 21 hours at 10 W/kg	Not clastogenic
ILS	human blood lymphocytes	chromosomal aberrations	1909.8 MHz in PCS mode 3 hours at 1.2, 2.9, 5.0, 10.1 W/kg and 21 hours at 10.2 W/kg	Not clastogenic
SRI	human blood lymphocytes	chromosomal aberrations	837 MHz in analog mode 3 hours at 1, 2.5, 5, 10 W/kg and 21 hours at 10 W/kg	Not clastogenic
ILS	human blood leukocytes	DNA damage in eukaryote cells	837 MHz in analog, TDMA mode 3 hours at 1, 2.5, 5, 10 W/kg and 24 hours at 10 W/kg	No DNA damage induced

* Integrated Laboratory Systems, Durham, NC. **SRI International, Menlo Park, CA.

Table 1 (continued on next page). Toxicological studies completed within the WTR program.

Lab	Biological model	Biological endpoint	Exposure conditions	Results
ILS*	human blood leukocytes	DNA damage in eukaryote cells	837 MHz in CDMA mode 3 hours at 1, 2.5, 5, 10 W/kg and 24 hours at 10 W/kg	There was an equivocal increase in DNA migration for a 3-hour exposure. No DNA damage was induced for the 24-hour exposure.
ILS	human blood leukocytes	DNA damage in eukaryote cells	1909.8 MHz in PCS mode 3 hours at 1.6, 2.9, 5.1, 10.3 W/kg and 24 hours at 10.1 W/kg	No DNA damage induced
ILS	human lymphocyte binucleate micronucleus cells	structural and/or numerical chromosomal damage; micronuclei formation	837 MHz in analog, TDMA, CDMA mode 3 hours at 1, 2.5, 5, 10 W/kg and 24 hours at 5, 10 W/kg	There was no increase in lymphocyte frequency for a 3-hour exposure. There was a statistically significant increase in lymphocyte frequency for a 24-hour exposure suggesting genotoxicity.
ILS	human lymphocyte binucleate micronucleus cells	structural and/or numerical chromosomal damage	1909.8 MHz in PCS mode 3 hours at 1.6, 2.9, 5.1, 10.1 W/kg and 24 hours at 10.3 W/kg	There was no increase in lymphocyte frequency for a 3-hour exposure. There was a statistically significant increase in lymphocyte frequency for a 24-hour exposure suggesting genotoxicity.
IN VIVO TOXICOLOGY				
ILS	rat brain cells	DNA damage in eukaryote cells	837 MHz in analog mode 3 hours at 1, 5, 10 W/kg	No statistically significant increase in DNA migration
UW*** (interim report)	rat brain cells	DNA damage in eukaryote cells	837 MHz in analog mode 3 hours at 1, 5, 10 W/kg	Additional testing with a larger sample size is needed before conclusion can be made.

* Integrated Laboratory Systems, Durham, NC. *** University of Washington, Seattle, WA.

Table 1 (continued from previous page). Toxicological studies completed within the WTR program.

Authors	Biological model	Biological endpoint	Exposure conditions	Results
IN VIVO TOXICOLOGY				
Malyapa R et al., 1998 U.S.	rat brain cells	DNA single-strand breaks	2450 MHz 2 hours 1.2 W/kg	No significant differences between groups in comet length or comet moment of cells; euthanasia by guillotine is most appropriate in studies of DNA damage in brain cells
Harris A et al., 1998 Australia	transgenic mice	malignant lymphoma induction	50 Hz 18 months	No tumorigenic effect
Imaida K et al., 1998 Japan	rats	rat liver carcinogenesis	929.2 MHz generated by TDMA 6 weeks 0.8-0.58 W/kg	No significant effect on rat liver carcinogenesis
Imaida K et al., 1998 Japan	rats	rat liver carcinogenesis	1.5 GHz generated by TDMA 6 weeks 0.680-0.453 W/kg	No significant effect on rat liver carcinogenesis
Frei M et al., 1998 U.S.	cancer-prone mice	latency, incidence, growth rate of mammary tumors, animal longevity	2450 MHz 18 months 0.3 W/kg	No significant effect on: mammary tumor incidence latency to tumor onset tumor growth rate animal longevity
Frei M et al., 1998 U.S.	cancer-prone mice	latency, incidence, growth rate of mammary tumors, animal longevity	2450 MHz 18 months 1.0 W/kg	No significant effect on: mammary tumor incidence latency to tumor onset tumor growth rate animal longevity

Table 2 (continued on next page). Toxicological studies identified through WTR literature surveillance.

Authors	Biological model	Biological endpoint	Exposure conditions	Results
Toler J et al., 1997 U.S.	cancer-prone mice	latency, incidence, growth rate of mammary tumors, animal longevity	435 MHz 21 months 0.32 W/kg	No significant effect on: mammary tumor incidence latency to tumor onset tumor growth rate animal longevity
Vijayalaxmi N et al., 1997 U.S.	cancer-prone mice	incidence of micronuclei in polychromatic erythrocytes (PCEs)	2450 MHz 18 months 1.0 W/kg	No statistically significant differences between RFR-exposed and sham-exposed mice in incidence of micronuclei in PCEs
Lai H & Singh N, 1997 U.S.	rat brain cells	DNA single-strand breaks	60 Hz 2 hours flux densities of 0.1, 0.25, 0.5 mT	A dose-dependent increase in single- and double-strand DNA breaks in rat brain cells (euthanasia by dry ice)
Lai H & Singh N, 1997 U.S.	rat brain cells treated with melatonin or a spin trap compound	blocking effects of RFR	2450 MHz 2 hours 1.2 W/kg	Melatonin and PBN blocked the RFR-induced increases in DNA breaks in rat brain cells
Repacholi M et al., 1997 Australia	transgenic mice	incidence of lymphoma	900 MHz 18 months 0.3-1.4 W/kg	Lymphoma risk found to be significantly higher in exposed group (OR=2.4, $p<0.006$)
Vollrath L et al., 1997 Germany	rats and hamsters	pineal melatonin synthesis	900 MHz up to 6 hours 0.06-0.36 W/kg--rats 0.04 W/kg--hamsters	No notable short-term effect on pineal melatonin synthesis
Fritze K et al., 1997 Germany	rats	blood-brain barrier permeability	900 MHz 4 hours 0.3-7.5 W/kg	No pathologically significant changes of the blood-brain barrier permeability
Singh B & Bate L, 1996 Canada	pigs	pulmonary intravascular macrophage (PIM) response	915 MHz (infrared, low and high microwave) 24 hours	Results suggest enhanced secretory activity in the PIMs of low microwave-treated pigs. High-microwave treatment induced some damage to pulmonary interstitium.

Table 2 (continued from previous page). Toxicological studies identified through WTR literature surveillance.

Authors	Biological model	Biological endpoint	Exposure conditions	Results
Balode Z, 1996 Latvia	cows living in the Skrunda RLS radiation zone	incidence of micronuclei in bovine peripheral erythrocytes	154-162 MHz impulse RFR exposure from a radio location station for at least 2 years	Statistically significant differences between the control and exposed groups in the frequency distribution of micronuclei in peripheral erythrocytes
Lai H & Singh N, 1995 U.S.	rat brain cells	DNA single-strand breaks	2450 MHz 2 hours 1.2 W/kg	An increase in single- and double-strand DNA breaks in rat brain cells (euthanasia by dry ice)
IN VITRO TOXICOLOGY				
Flipo D et al., 1998 Canada	murine macrophages, spleen lymphocytes, and thymic cells of mice	alteration of cellular functions of immunocompetent cells	250-1500 G static magnetic fields 24 hours	Significant effects recorded: decreased macrophage phagocytosis inhibited lymphocyte mitogenic response enhanced apoptotic death of thymocytes
Ivaschuk O et al., 1997 U.S.	nerve growth factor-treated PC12 rat pheochromocytoma cells	effects on c-*jun* and c-*fos* expression	836.55 MHz generated by TDMA up to 60 minutes	No change in c-*fos* transcript levels. Transcript levels for c-*jun* altered only after 20-minute exposure to 9 mW/cm^2
Maes A et al., 1997 Belgium	human whole blood cells	chromosome aberrations, sister chromatid exchanges, DNA single-strand breaks	935.2 MHz (GSM) generated by CDMA alone and in combination with mitomycin C 2 hours 0.3-0.4 W/kg	No direct cytogenic effect was found (alkaline comet assay was used)
Vijayalaxmi N et al., 1997 U.S.	human blood lymphocytes	incidence of chromosomal aberrations and micronuclei	2450 MHz 90 minutes 12.46 W/kg	No significant differences between RFR-exposed and sham-exposed lymphocytes with regard to mitotic indices, incidence of cells showing chromosomal damage, exchange aberrations, acentric fragments, binucleate lymphocytes, micronuclei

Table 2 (continued from previous page). Toxicological studies identified through WTR literature surveillance.

Authors	Biological model	Biological endpoint	Exposure conditions	Results
Antonopoulos A et al., 1997 Germany	human peripheral lymphocytes	cell-cycle progression & frequencies of sister chromatid exchanges	380 MHz at 0.08 W/kg 900 MHz at 0.208 W/kg 1800 MHz at 1.7 W/kg	No differences between treated and control cultures were found
Stagg R et al., 1997 U.S.	rat glioma cells	DNA synthesis rates	836.55 MHz generated by TDMA 24 hours 0.15 - 59 uW/g	No increase in cell proliferation
Cain C et al., 1997 U.S.	10T I/L cells and 10e cells	intercellular communication disruption	836.55 MHz generated by TDMA 28 days 0.15, 1.5, 15 uW/g	No influence on tumor promotion
Malyapa R et al., 1997 U.S.	human glioblastoma and mouse cells	DNA damage as measured by alkaline comet assay	2450 MHz up to 24 hours 0.7 and 1.9 W/kg	No DNA damage in mammalian cells
Malyapa R et al., 1997 U.S.	human glioblastoma and mouse cells	DNA damage as measured by alkaline comet assay	835.62 FMCW & 847.74 MHz generated by CDMA 24 hours 0.6 W/kg	No DNA damage in mammalian cells
Garaj-Vrhovac V et al., 1996 Croatia	human blood lymphocytes	frequency of micronuclei	415 MHz up to 30 minutes	Increased frequency of micronuclei
Cleary S et al., 1996 U.S.	cytolytic T lymphocyte (in the absence or presence of interleukin-2)	lymphocyte proliferation	2450 MHz 2 hours 5-50 W/kg	RF radiation may affect membrane signal transduction in the absence of heating, resulting in altered proliferation
Wolke S et al., 1996 Germany	cardiac myocytes of guinea pigs	calcium homeostasis of isolated ventricular cardiac myocytes	900 MHz 1300 MHz 1800 MHz 500 seconds	No significant differences between sham exposure and high-frequency field exposure of cells

Table 2 (continued from previous page). Toxicological studies identified through WTR literature surveillance.

Authors	Biological endpoint	Exposure conditions	Results
Brusick D et al., 1998 U.S.	genotoxicity of RFR	800-3000 MHz RFR	Data from over 100 studies were subjected to a weight-of-evidence evaluation and suggest that RFR is not directly mutagenic and that adverse effects from organism exposure to RFR are predominantly the result of hypothermia.
Verschaeve L & Maes A, 1998 Belgium	genetic toxicology of RFR	mobile telephone frequencies	Majority of research shows that such exposure does not induce genetic effects in vivo or in vitro and does not seem to be teratogenic or to induce cancer, however, more research should be done.
Hermann D & Hossmann K, 1997 Germany	neurological effects	microwave exposure	No positive evidence that pulsed or continuous microwave exposure confers elevated risk to brain.

Table 3. Toxicological reviews identified through WTR literature surveillance.

III

IN VITRO STUDIES/ GENOTOXICOLOGY

8

STATE OF THE SCIENCE: RF GENETIC TOXICOLOGY

J. Morrissey

The use of cellular telephones has increased rapidly since they became generally available to the public around the year 1989. Currently, over 200 million people in the world subscribe to wireless communication services, and this number is expected to grow at a fast pace until well past the turn of the century. In parts of the world with less developed infrastructure, such as regions of Latin America, Africa, and Eastern & Far Eastern countries, wireless communications may soon become the primary means of telephonic communication.

With the rapid growth of wireless technology, however, has come an element of public concern over potential human health effects from long-term exposure to low-level radiofrequency (RF) radiation. This concern originated in 1992 when an anecdotal report of brain cancer developing in a St. Petersburg, FL, cellular telephone user received national attention. Since then, worry that cellular telephony might be linked to health disorders including cancer, headaches, sleep disturbances, and memory loss has been raised in the U.S. and throughout the world. The media are quick to report on claims of health effects related to the use of cellular telephones, but unfortunately they often misinterpret study results and provide more confusion than factual information to consumers. In contrast, a number of recognized scientific and standard-setting bodies have made more comprehensive reviews of the available literature and concluded that no health hazard is apparent when wireless equipment is designed, manufactured, and operated in conformance with internationally-accepted exposure limits.

Even though no health hazard has been linked to the use of cellular phones, government and industry have recognized the need to look into the matter further and address the public's concerns. In this regard, almost 150 studies have been initiated to examine the effects of wireless communication-specific RF exposure on biological systems (1). Current industry sponsors of this research include Motorola, Ericsson, Nokia, Alcatel, Deutsche Telekom, France Telecom, NTT DoCoMo, Belgacom, Swiss Telecom, and Telstra. In addition, research has been sponsored through four industry-funded groups: Wireless Technology Research (WTR, U.S.), Forschungsgemeinschaft Funk (FGF, Germany), Mobil Tele Branshcen (MMT, Scandinavia), and the Association for Radio Industries & Businesses (ARIB, Japan), as well as by the Australian, Japanese, Finnish, French, Italian, Hungarian, Polish, Swedish, and German governments. A list of ongoing and recently completed studies can be found on the World Health Organization (WHO) website (http://www-nt.who.int/peh-emf/database.htm).

In an effort to coordinate future research efforts in this area, WHO has also developed an agenda of research items which will best augment the existing database of studies and provide the most useful information for health risk assessment. Using this agenda as a guide, additional research efforts have been proposed through the European Commission's 5th Framework Program including animal studies, in vitro studies, and human studies, as well as a large multi-center epidemiological study proposed by the International Agency for Research on Cancer (IARC). The data generated by these studies will allow public health agencies and standard-setting bodies to make appropriate risk evaluations. Such an evaluation to determine cancer risk from cellular telephone RF exposure will be performed by IARC in June of 2003. A similar evaluation by WHO to determine human non-cancer disease risk is scheduled for 2004.

An important part of the risk analysis database includes studies to determine whether RF exposure is genotoxic. A total of 105 studies of this type (65 in vitro, 28 in vivo, and 12 drosophila studies) using exposures between 800 MHz and 3 GHz have been reviewed by Brusick et al. (2). Upon review, 73% of these studies were found to be negative. Further, no positive study has yet been validated by independent replication, and many have been directly contradicted by other reports. The authors conclude that "[when all studies are considered together, non-thermal levels of] RF exposure do not appear to be directly mutagenic or genotoxic." Although the review included four in vitro studies of micronuclei formation all reporting positive findings, only three of the four were relevant to human exposure and these were performed at relatively high exposure levels in exposure systems that were not well characterized. Whether these conditions lead to the introduction of heating artifacts is difficult to ascertain. In the one human study reviewed by Brusick et al. showing positive micronuclei formation, possible competing (confounding) exposures make it difficult to interpret the results.

Relevant studies published since the Brusick et al. review largely support the conclusion that RF is not genotoxic, although some positive findings have been reported which do need careful and immediate clarification. Data presented at the June 1999 WTR meeting in Long Beach by Drs. Hook and Tice of Integrated Laboratory Systems (ILS) show an increased rate of micronuclei formation in human lymphocytes following a 24-hour exposure to CDMA, TDMA, PCS, or FM signals at an average specific absorption rate (SAR) of 10 W/kg (3). Peak SAR values within the exposure system, however, were calculated to be as much as double the average value using numerical models. Some effects were also observed at 5 W/kg (average SAR). In contrast, preliminary studies from the Roti Roti group at Washington University showed no increase in micronuclei formation following extended exposures to CDMA or FM signals at 5 W/kg, although these studies were performed in cell lines and using a different exposure system (4). Three recently reported studies funded by the FGF also showed no effects of RF signal exposure on micronuclei formation or on any other type of chromosome aberration in human lymphocytes or yeast (5-7).

In recent studies of DNA breaks, damage, and mutation, the majority of data point again to a lack of an effect by RF exposure. While Phillips et al. (8) and Lai & Singh (9-11) have both reported changes in the level of DNA breaks in cells or cell lines following RF exposure, the observed effects between the two groups are difficult to correlate as they appear to move in opposite directions. These studies are further contrasted by recent reports from the Roti Roti group at Washington University (12-14) and by Hook and Tice of ILS (3) showing no effect of RF exposure on the level of DNA breaks using a number of cell phone signal types and exposure levels. While a recent study by Maes et al. (15) using low-level RF (GSM) exposure duplicated their earlier observation of an effect on sister chromatid exchange (SCE) rate, this study is contrasted by a number of earlier reports of no effect on SCE. Further, in the recent Maes et al. study there was no observation of increased micronuclei formation or DNA breaks.

In conclusion, the available data continue to support the conclusion that RF exposure of the type and level emitted by cellular telephones does not appear to be genotoxic. There are, however, isolated reports of positive effects. While these positive findings should certainly not be discounted, these findings must be carefully and independently replicated by another laboratory before they can be considered as a genuine and non-thermal effect of RF exposure.

References

1. World Health Organization (WHO) Electromagnetic Field (EMF) database at http://www-nt.who.int/peh-emf/database.htm.

2. Brusick D, Albertini R, McRee D, Peterson D, Williams G, Hanawalt P, & Preston J. Genotoxicity of radiofrequency radiation. Environ Mol Mutagen 1998; 32(1): 1-16.

3. Hook GJ. Evaluation of potential genotoxicity of radiofrequency radiation technologies using the single cell gel electrophoresis and micronucleus assays. In: GL Carlo, editor. Wireless phones and health II: State of the science. Norwell, MA: Kluwer Academic Publishers; 2000.

4. Roti Roti JL, Malyapa RS, Higashikubo R, Zhang P, Li L, Straube WL, Moros EG, & Pickard WF. The carcinogenic potential of 835.62 MHz FMCW and 847.74 MHz CDMA radiations: Ongoing studies. In: GL Carlo, editor. Wireless phones and health II: State of the science. Norwell, MA: Kluwer Academic Publishers; 2000.

5. Eberle P. Electromagnetic compatibility of biological systems. In: Brinkmann G & Friedrich G, editors. Berlin: VDE-Verlag; 1997; Eberle P. Cytogenic studies of the effects of mobile telephone radio waves. Newsletter Edition Wissenschaft 1996 Feb; 4(1): 1-15.

6. Gos P. Mobile phone type electromagnetic fields do not influence genetic stability in yeast.

Second World Congress for Electricity and Magnetism in Biology and Medicine. 1997 June 8-13; Bologna, Italy. 1997.

7. Antonopoulos A. Effects of high-frequency electromagnetic fields on human lymphocytes in vitro. Mutat Res 1997; 395(2-3): 209-214.

8. Phillips JL, Ivaschuk O, Ishida-Jones T, Jones RA, Campbell-Beachler M, & Haggren W. DNA damage in molt-4 lymphoblastoid cells exposed to cellular telephone radiofrequency fields in vitro. Bioelectrochem Bioenerg 1998; 45(1): 103-110.

9. Lai H & Singh NP. Acute low-intensity microwave exposure increases DNA single-strand breaks in rat brain cells. Bioelectromagnetics 1995; 16(3): 207-210.

10. Lai H & Singh NP. Single- and double-strand DNA breaks in rat brain cells after acute exposure to radiofrequency electromagnetic radiation. Int J Radiat Biol 1996; 69(4): 513-521.

11. Lai H & Singh NP. Melatonin and a spin-trap compound block radiofrequency electromagnetic radiation-induced DNA strand breaks in rat brain cells. Bioelectromagnetics 1997; 18(6): 446-454.

12. Malyapa RS, Ahern EW, Bi C, Straube WL, LaRegina M, Pickard WF, & Roti Roti JL. DNA damage in rat brain cells after in vivo exposure to 2450 MHz electromagnetic radiation and various methods of euthanasia. Radiat Res 1998; 149(6): 637-645.

13. Malyapa RS, Ahern EW, Straube WL, Moros EG, Pickard WF, & Roti Roti JL. Measurement of DNA damage after exposure to 2450 MHz electromagnetic radiation. Radiat Res 1997; 148(6): 608-617.

14. Malyapa RS, Ahern EW, Straube WL, Moros EG, Pickard WF, & Roti Roti JL. Measurement of DNA damage after exposure to electromagnetic radiation in the cellular phone communication frequency band (835.62 and 847.74 MHz). Radiat Res 1997; 148(6): 618-627.

15. Maes A, Collier M, Van Gorp U, Vandoninck S, & Verschaeve L. Cytogenic effects of 935.2 MHz (GSM) microwaves alone and in combination with mitomycin C. Mutat Res 1997; 393(1-2): 151-156.

9

EVALUATION OF POTENTIAL GENOTOXICITY OF RADIOFREQUENCY RADIATION TECHNOLOGIES USING A BATTERY OF IN VITRO TESTS

G.J. Hook

This paper presents an overview of the results from a battery of in vitro bacterial and mammalian cell assays used to evaluate the genotoxicity of radiofrequency (RF) fields produced by cellular communication technologies. The work was sponsored by Wireless Technology Research, LLC (WTR) and conducted at Integrated Laboratory Systems (ILS). The following individuals participated in the conduct of these studies: Dr. Maria Donner, Dr. Raymond Tice, Ms. Marie Vasquez, Mr. Joe Clancy, Ms. Diane Satterfield, Ms. D. Blackwell, Ms. Lori Philips, and Dr. Scott Clark.

Exposure System

The exposure system used in these tests is described in an accompanying chapter (see Chapter 2 in this volume). The exposure system consisted of two transverse electromagnetic (TEM) cells set up in series. This allowed for the application of dose rates at the same time. Each experiment consisted of a sham-exposed control group, a positive (chemical) control group, and four different specific absorption rate (SAR) groups. The SARs used in this study were 1, 2.5, 5, and 10 W/kg. Cells were exposed in test tubes set up three on each side of a septum of the TEM. Temperature was controlled using a circulating water jacket, internal fans, and water baths to control the temperature of the circulating water.

The culture temperatures were monitored using fiber optic probes, placed through the door into the test tubes. They were placed in two positions: one in the middle of the test tube and one at the bottom. Those cells were not used as part of the experiment. Temperature was maintained at 37±1°C and recorded throughout the exposure.

For these experiments, a variety of methods were used to generate the RF signal. In all cases the signal was amplified using a Calmus L500G power amplifier. For the analog studies a signal generator was used and the signal was voice modulated using a CD recoding supplied by Motorola. For the digital TDMA and CDMA and the PCS experiment, cellular telephones were used to generate signals. These were supplied by Motorola and adapted for our use by them. TDMA was voice modulated; CDMA was not voice modulated. The PCS signal was at 1909.8 MHz using GSM type modulation and was not voice modulated.

Test Battery

Three different tests were used: a bacterial mutagenesis test, the *Salmonella/E. coli* (AMES) assay; a mammalian mutagenesis test, the mouse lymphoma Tk± (MOLY) assay; and a clastogenicity assay, the human blood lymphocyte (HULY) assay.

In each experiment, except for the extended-exposure HULY assay, exposures were for 3 hours. In the HULY assay, one experiment used a 3-hour exposure and the other a 21-hour exposure. In all experiments concurrent positive and sham negative controls were included. In all experiments the positive controls showed a significant increase over the concurrent sham exposed controls. In each assay a measure of toxicity was included; there was no evidence of toxicity in any experiment using any assay.

AMES Assay

This assay used five bacterial strains: four *Salmonella* strains and one *E. coli* strain. A modified pre-incubation assay was used, where the cells were exposed before plating. In all cases there were strain-specific direct acting positive controls included. Following the exposure the bacteria were plated on supplemented agar plates and scored for the number of revertant (mutant) colonies using an automatic colony counter.

MOLY Assay

For the MOLY assay the positive control was ethylmethane sulfanate. The cells were washed, cultured overnight prior to exposure, set up in tubes, and exposed. To evaluate the induced mutant frequency three plates were prepared per exposed culture tube. In each experiment there were two culture tubes. Following exposure the cells were incubated at 37°C for 10-12 days, and then the number of mutant colonies were counted using an automatic colony counter.

HULY assay

Human peripheral blood from a single donor was used for each experiment. For the HULY assay the positive control was mitomycin C. Whole blood was put into a

Technology	Assay	Result	Conclusion
ANALOG	AMES	Negative	Not mutagenic
		Equivocal TA1535, TA98	
	MOLY	Negative	Not mutagenic
		Negative	
	HULY	Negative	No induction of chromosome aberrations or polyploidy
		Negative	
TDMA	AMES	Negative	Not mutagenic
		Negative	
	MOLY	Negative	Not mutagenic
		Negative	
	HULY	Negative	No induction of chromosome aberrations or polyploidy
		Negative	
CDMA	AMES	Equivocal TA1535	Not mutagenic
		Negative	
	MOLY	Equivocal	Not mutagenic
		Negative	
	HULY	Negative	No induction of chromosome aberrations or polyploidy
		Negative	
PCS	AMES	Negative	Not mutagenic
		Negative	
	MOLY	Negative	Not mutagenic
		Negative	
	HULY	Negative	No induction of chromosome aberrations or polyploidy
		Negative	

Table 1. Test battery results.

culture media and stimulated to divide using phytohemagluttanin. Following a 48-hour culture period, duplicate cell cultures were transferred to exposure tubes. For the 3-hour exposure test, cells were cultured for 21 hours following exposure. For the extended exposure test, cells were exposed for 21 hours then cultured for 2 hours in an incubator. Colchicine was added for the last 2 hours of culture for both tests to arrest the cells in metaphase. For each experiment 100 metaphase spreads with 46±1 chromosomes from each of two replicate cultures were scored.

Summary of Results

A summary of all the experiments is presented in Table 1. Overall RF fields generated by cellular technologies or at cellular telephone frequencies were not mutagenic in bacterial or mammalian systems. In addition there was no evidence of an induction of chromosome aberrations or polyploidy. The lack of genotoxic effects was observed for both digital and analog technologies, with and without voice modulation, and at different frequencies (837 and 1909.8 MHz).

10

MEASUREMENT OF ALKALI LABILE DNA DAMAGE AND CROSS-LINKS FOLLOWING 2450 MHZ MICROWAVE AND LOW DOSE GAMMA IRRADIATION IN VITRO

I. Lagroye, B. Wettring, E.G. Moros, W.L. Straube
W.F. Pickard, & J.L. Roti Roti

Introduction

DNA damage is thought to be a contributor to and/or an initiator of the neoplastic process. Therefore, report of detectable DNA damage being present in rat brain after whole-body exposure to 2450 MHz microwaves (Lai & Singh 1995, 1996) is of potential concern. If verified, a plausible mechanism for carcinogenesis would have been supported. However, initial attempts to confirm the above results have failed to detect measurable DNA damage in cultured cells and in brain cells irradiated in vivo (Malyapa et al. 1997, 1998a). However, there are potentially significant differences in the methods used in the studies that gave conflicting results.

These studies were undertaken to resolve two possible reasons for the discrepancy between the reported results of Lai & Singh (1995, 1996) and those of Malyapa et al. (1998a). Specifically, these issues relate to assay sensitivity and the ability to detect DNA-protein or DNA-DNA cross-links, referred here as DNA cross-links. In regard to the former, studies have been published showing that the method used according to Olive et al. (1992) can detect DNA damage following irradiation with 1 cGy or lower (Malyapa et al. 1998b). Therefore, we sought to determine if an independent team could detect DNA damage following similar doses of ionizing radiation. The second issue was that the method used by Malyapa et al. (1997, 1998a, 1998b) could not detect the presence of DNA cross-links. Therefore, we combined 2450 MHz microwave exposure with 4 Gy of gamma irradiation to determine if the microwave exposure induced DNA cross-links. As a positive control for such cross-links, we also treated the cells with Cis-DiammineDichloroPlatinum II (CDDP), an agent known to induce both DNA-DNA and DNA-protein cross-links (Ducore 1986; Samuel et al. 1998). Positive results with the cross-linking assay could have resolved the differences between the results of Lai & Singh and those of Malyapa et al.

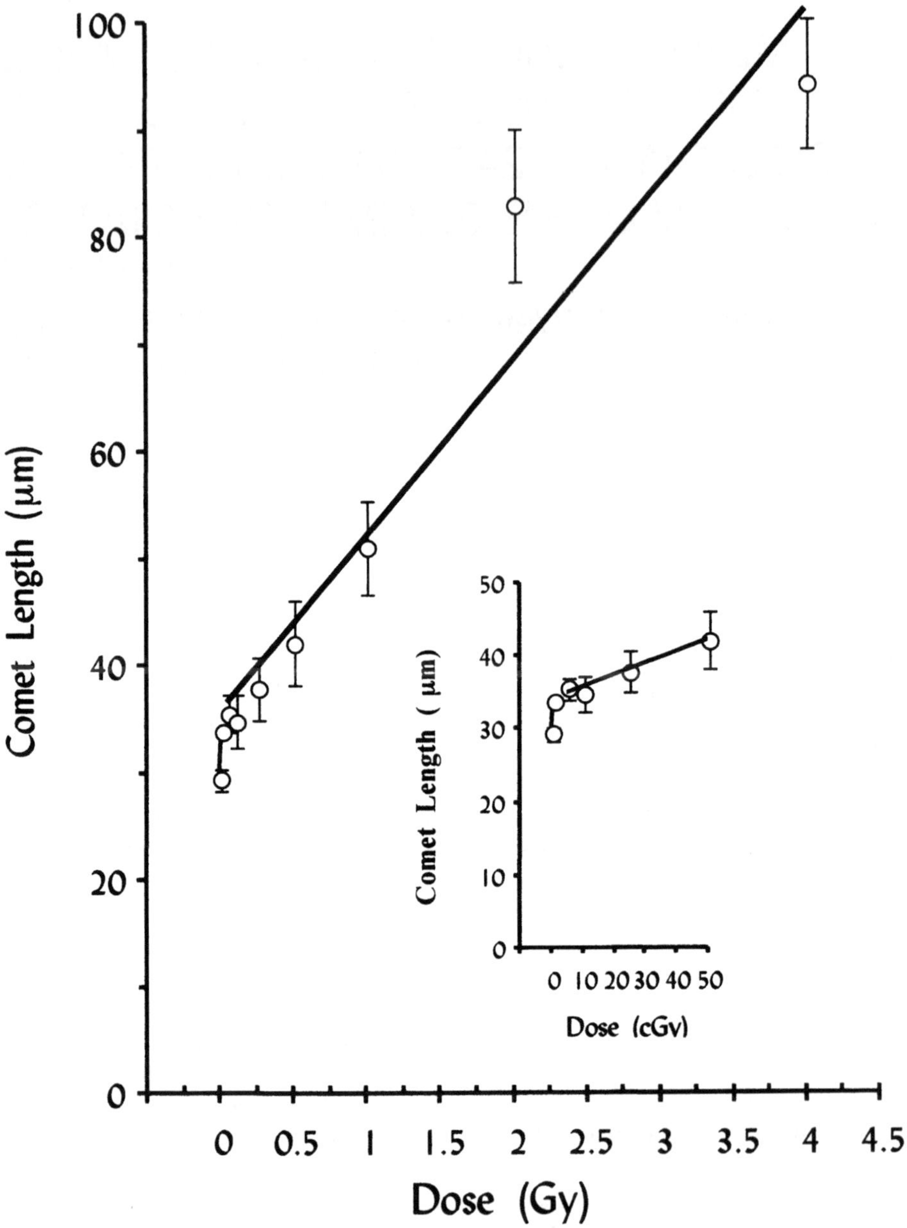

Figure 1. **Dose-response curve for the comet length (in µm) of C3H 10T1/2 cells after gamma radiation. Results are expressed as the mean ± SEM of 3 to 11 independent experiments. The enclosed panel shows the dose-response curve obtained in the low dose-range (<50 cGy) of ionizing radiation.**

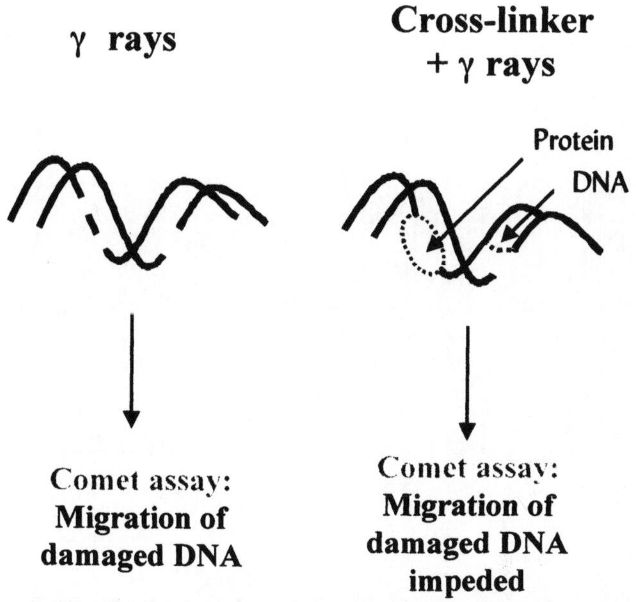

Figure 2. DNA damage and DNA cross-linking illustrative assay. In this cartoon, the presence of DNA breaks after gamma radiation is evidenced as the damaged DNA migrate using the alkaline comet assay (left panel), whereas using the same assay the DNA cross-links are detected as the DNA migration induced by ionizing radiation is obliterated (right panel).

In order to specifically address these issues, we conducted these studies using cultured, non-transformed cells, namely the C3H 10T1/2 cell line. This approach has the advantage of avoiding the expense and complications of in vivo studies. Also, a better defined and more homogenous specific absorption rate (SAR) could be obtained by using cultured cells as opposed to exposure in the Guy-type irradiator (Moros et al. 1999). These studies were conducted using a SAR of 1.9 W/kg, the highest SAR used in the previous studies.

Low Ionizing Radiation Dose Sensitivity

To determine the sensitivity of the comet assay for gamma (γ) ray-induced DNA damage, exponentially growing C3H 10 T1/2 cells were irradiated in suspension and on ice in order to prevent repair processes. The following doses of γ radiation were used: 0.01, 0.05, 0.1, 0.25, 0.5, 1 and 4 Gy. According to the dose delivered, two dose rates were used: 0.1 Gy/min from 0.01 to 0.1 Gy and 0.8 Gy/min from 0.25 to 4 Gy. Results are expressed as the mean comet length (in μm) ± SEM of at least three independent experiments. The nonparametric Mann-Whitney U-test (unpaired, one-tailed) was used to determine statistical significance. As shown in Figure 1, the comet length increased with the dose of γ rays. As previously reported (Malyapa et al. 1998), the dose-response curve showed two slopes: one from 0 to 1 cGy and

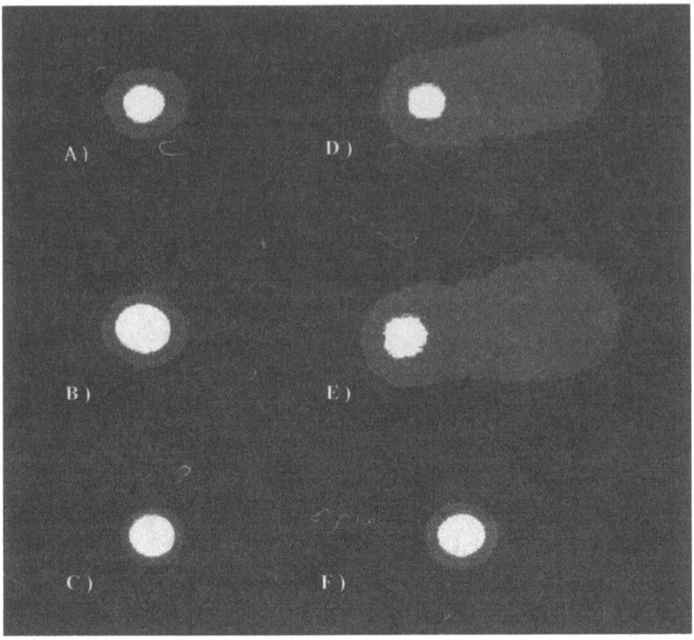

Figure 3. Representative comet images of the six groups of samples: A) Sham exposure; B) 2-hour exposure to 2450 MHz MW; C) 2-hour exposure to CDDP (1 mM); D) 4-Gy (137 Cs) irradiation; E) 4-Gy irradiation after a 2-hour pretreatment with 2450 MHz MW; F) 4-Gy irradiation after a 2-hour pretreatment with CDDP (1 mM). Images were digitized using the Optimas® software.

another in the higher range (> 1 cGy). The lowest dose of γ rays we used, 1 cGy, was able to induce significant DNA damage as measured by the comet length. Thus, we confirmed the sensitivity of the version of the alkaline comet assay used in our laboratory, as well as the shape of the dose-response curve.

Damage after 2450 MHz

Both alkaline labile DNA damage and DNA cross-links were measured using the alkaline comet assay (Olive et al. 1992); the comet length (in μm) and normalized comet moment (Kent et al. 1995) were the parameters studied. The procedure for DNA cross-links was adapted from Pfülher and Wolf (1996), using γ rays as a genotoxic agent (Figure 2). Briefly, cross-links are evidenced as a reduction in the migration of damaged DNA induced by a genotoxic agent. As stated before, the chemical CDDP (1 mM, 2 hours) was used as a positive control for DNA cross-links. Treatment with 1 mM CDDP greatly impeded DNA migration induced by 4 Gy of γ rays as shown in Figure 3.

Exponentially growing C3H 10T1/2 cells were exposed to 2450 MHz MW for 2 hours. A Radial Transmission Line (RTL) system was used to expose the cells as

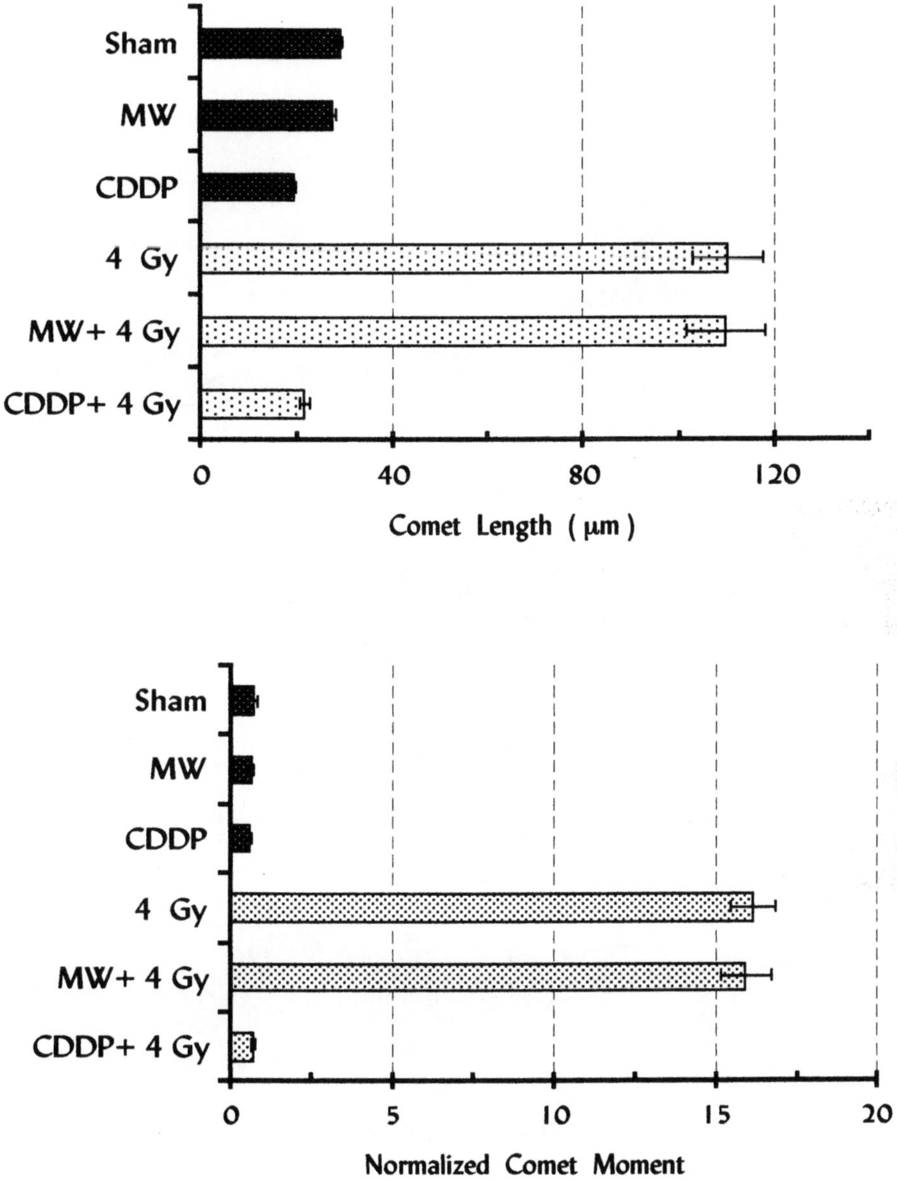

Figure 4. Mean comet length (upper panel) and mean comet moment (lower panel) of sham-, 2450 MHz MW-, CDDP (1mM)-, 4 Gy-, CDDP + 4 Gy-, and 2450 MHz MW + 4 Gy-treated samples.

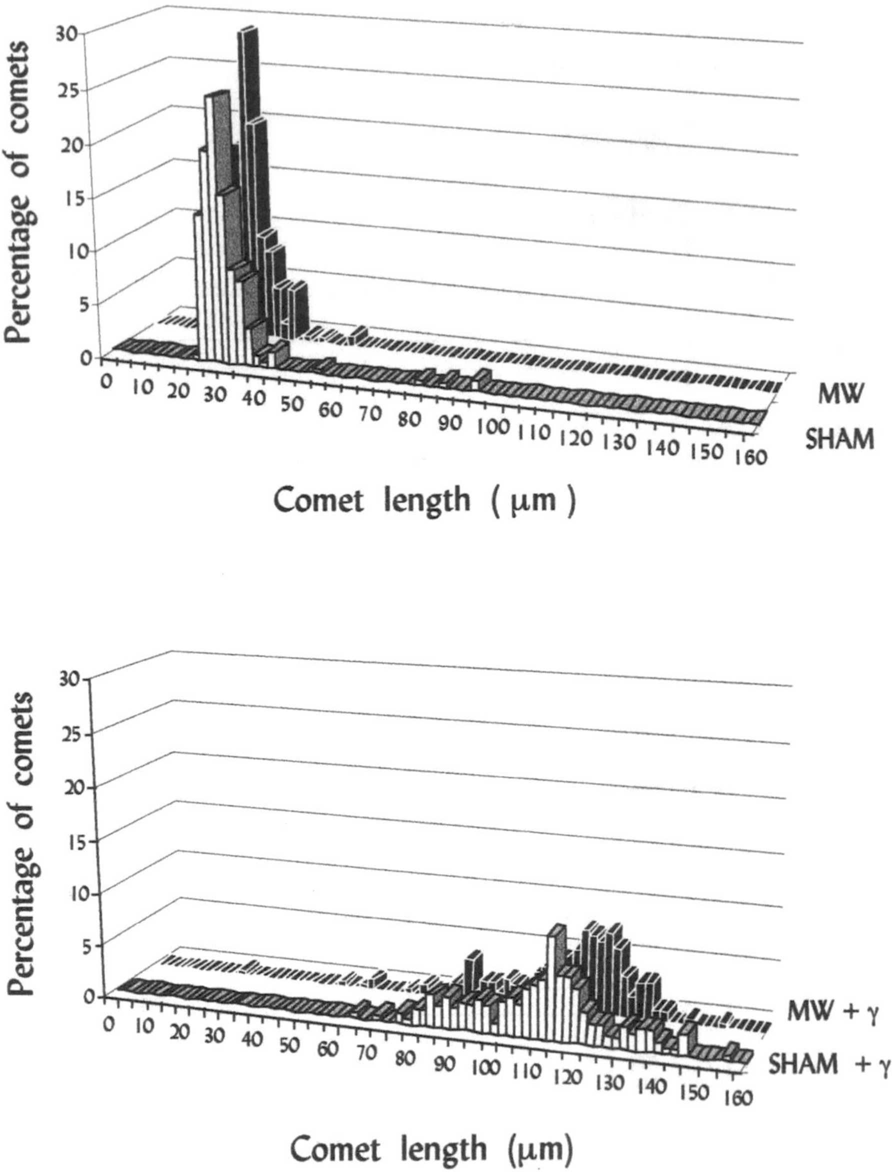

Figure 5. Percent distribution of C3H 10T1/2 cells as a function of comet length in μm. Upper panel: Distribution of sham- and 2450 MHz microwave-exposed samples. Lower panel: Distribution of samples pretreated with a 2-hour sham- or 2450 MHz MW-exposure before being irradiated with 4 Gy of gamma rays. Three to four experiments were pooled.

previously described (Moros et al. 1999). Cells were then irradiated or mock-irradiated with 4 Gy of γ rays, as described above. All steps were performed in the dark. Experimental procedures for the alkaline comet assay and for image analysis were performed as described in Malyapa et al. (1997, 1998). The unpaired, two-tailed Student's t test was used to determine statistical significance. As shown in Figure 4 (upper three bars, both panels), the mean comet length but not the mean comet moment was significantly reduced after exposure to 1 mM CDDP for 2 hours ($p<0.0001$ and $p=0.4$, respectively). Thus, comet length could detect DNA cross-links induced by CDDP in the absence of a consecutive γ irradiation. The inability for normalized comet moment to detect DNA cross-links could be due to DNA-loop conformation in dividing cells or to a lack of dynamic range in the comet moment parameter. In contrast, the differences in both parameters between sham-exposed cells and cells exposed to 2450 MHz MW were not significantly different ($p=0.14$ and $p=0.26$ for the comet length and the comet moment, respectively). Significant DNA migration (due to DNA damage) was detected after 4 Gy of ionizing radiation (Figure 4, three lower bars, both panels). A 2-hour pretreatment with 1 mM CDDP completely impeded the migration induced by 4 Gy of γ rays. However, prior exposure to 2450 MHz CW microwaves at 1.9 W/kg had no effect on this migration, showing an absence of DNA cross-linking. The absence of any significant alteration in DNA migration can be seen in the comet length histogram shown in Figure 5. These results demonstrate that exposure to 2450 MHz MW at a SAR of 1.9 W/kg did not induce DNA cross-linking.

Conclusions

As used in our laboratory, the alkaline comet assay is able to detect the alkaline labile DNA damage induced by doses of ionizing radiation as low as 1 cGy and DNA cross-links induced by 1 mM of CDDP. However, neither DNA damage nor DNA cross-links could be detected after a 2-hour exposure to 2450 MHz CW microwave at a SAR of 1.9 W/kg. Thus the difference between our results and those of Lai and Singh cannot be due to differences in the sensitivity of the assay or the ability to detect DNA cross-links.

Acknowledgment

Motorola Corporation supported this work.

References

Ducore JM. Cis-Diamminedichloroplatinum(II) (CDDP)-induced crosslinking and crosslink removal in L1210 cells in vitro after theophylline co-treatment. Biochem Pharmacol 1986; 35: 607-611.

Kent CRH, Eady JJ, Ross GM, & Steel GC. The comet moment as a measure of DNA damage in the comet assay. Int J Radiat Biol 1995; 67: 655-660.

Lai H & Singh NP. Acute low-intensity microwave exposure increases DNA single-strand breaks in rat brain cells. Bioelectromagnetics 1995; 16: 207-210.

Lai H & Singh NP. Single- and double-strand DNA breaks in rat brain cells after acute exposure to radiofrequency electromagnetic radiation. Int J Radiat Biol 1996; 69: 513-521.

Malyapa RS, Ahern EW, Bi C, Straube WL, LaRegina M, Pickard WF, & Roti Roti JL. DNA damage in rat brain cells after in vivo exposure to 2450 MHz electromagnetic radiation and various methods of euthanasia. Radiat Res 1998a; 149: 637-645.

Malyapa RS, Ahern EW, Straube WL, Moros EG, Pickard WF, & Roti Roti JL. Measurement of DNA damage after exposure to 2450 MHz electromagnetic radiation. Radiat Res 1997; 148: 608-617.

Malyapa RS, Bi C, Ahern EW, & Roti Roti JL. Detection of DNA damage by the alkaline comet assay following low dose ionizing radiation. Radiat Res 1998b; 149: 396-400.

Moros EG, Straube WL, & Pickard WF. The radial transmission line as a broad-band shielded exposure system for microwave irradiation of large number of flasks. Bioelectromagnetics 1999; 20: 65-80.

Olive PL, Wlodek D, Durand RE, & Banath JP. Factors influencing DNA migration from individual cells subjected to gel electrophoresis. Exp Cell Res 1992; 198: 259-267.

Pflülher S & Wolf HU. Detection of DNA-crosslinking agents with the alkaline comet assay. Environ Mol Mutagenesis 1996; 27: 196-201.

Samuel SK, Spencer VA, Bajno L, Sun JM, Holth LT, Oesterreich S, & Davie JR. In situ cross-linking by cisplatin of nuclear matrix-bound transcription factors to nuclear DNA of human breast cancer cells. Cancer Res 1998; 58: 3004-3008.

11

EVALUATION OF POTENTIAL GENOTOXICITY OF RADIOFREQUENCY RADIATION TECHNOLOGIES USING THE SINGLE CELL GEL ELECTROPHORESIS AND MICRONUCLEUS ASSAYS

G.J. Hook

This paper presents work performed at Integrated Laboratory Systems (ILS) and supported by Wireless Technology Research, LLC (WTR). In these experiments, the single cell gel (SCG) and the micronucleus assays were used to determine genotoxicity. For both assays human peripheral blood lymphocytes were used, and the same set of cultures was used to look at both of those endpoints. The following individuals participated in the conduct of these studies: Dr. Maria Donner, Dr. Raymond Tice, and Ms. Marie Vasquez.

Exposure System

The exposure system used in these tests is described in an accompanying chapter (see Chapter 2 in this volume). The exposure system consisted of two transverse electromagnetic (TEM) cells set up in series. This allowed for the application of dose rates at the same time. Each experiment consisted of a sham-exposed control group, a postive (chemical) control group, and four different specific absorption rate (SAR) groups. The SARs used in this study were 1, 2.5, 5, and 10 W/kg for 3-hour exposures. For the 24-hour exposures only one SAR was used at a time. For analog and TDMA simulation frequencies, studies were conducted using 5 or 10 W/kg. For CDMA and PCS simulation frequencies, studies were conducted using either 1 or 10 W/kg. Cells were exposed in test tubes set up three on each side of a septum of the TEM. Temperature was controlled using a circulating water jacket, internal fans, and water baths to control the temperature of the circulating water.

The culture temperatures were monitored using fiber optic probes, placed through the door into the test tubes. They were placed in two positions: one in the middle of the test tube and one at the bottom. Those cells were not used as part of the experiment. Temperature was maintained at 37±1°C and recorded throughout exposure.

For these experiments, a variety of methods were used to generate the RF signal. In all cases the signal was amplified using a Calmus L500G power amplifier. For the analog studies a signal generator was used and the signal was voice modulated using a CD recoding supplied by Motorola. For the digital TDMA and CDMA and the PCS experiment, cellular telephones were used to generate signals. These were supplied by Motorola and adapted for our use by them. TDMA was voice modulated; CDMA was not voice modulated. The PCS signal was at 1909.8 MHz using GSM type modulation and was not voice modulated.

Test System

The test system used unstimulated (non-dividing) human lymphocytes. The cells were exposed to radiofrequency (RF) fields at SARs of 1.0, 2.5, 5.0, and 10.0 W/kg. After RF exposure, the cells were split into two. A sample of the one set of cells were put onto clear slides, lysed, treated with proteinase K, put into an alkaline media to allow for DNA unwinding, and then electrophoresed. The slides were neutralized, dehydrated with ethanol, stained with Syber Green™, and analyzed using the Komet 3.1 image analyzer. In addition to the SCG assay itself, there were two measures of toxicity: evaluation of the number of cells with low molecular weight DNA and measurement of cell viability using a dual dye exclusion assay. No evidence of toxicity was observed in any of these experiments using either assay.

Three measures were used to evaluate the effect using the SCG assay. One was the tail length which is the measure of the length of the comet tail. The second measure was the percent of DNA that was in the tail, and the third measure was the tail moment which is the integration of these two measures.

For the micronucleus assay, following exposure, the cells were put into a medium containing phytohemagluttanin (PHA) to stimulate them to divide. The cells were then placed into an incubator at 37±1°C. Cytochlasin B was added 44 hours after initiation with PHA. Cytochlasin B inhibits cytokinesis so that cells that go through one and only one division will have two main nuclei. The number of binucleate cells with micronuclei was used to evaluate the clastogenicity of RF fields. At 72 hours, the cells were treated with hypotonic, stained, air dried, and scored. The measures of toxicity for this assay were the binucleate cell index (the number of binucleate cells per 1000 cells) and the replicative index (a measure of how many divisions the cells have gone through). In no case was evidence of toxicity due to exposure to RF fields observed using either toxicity assay.

Concurrent positive controls (ethylmethane sulfonate-treated cells) were included in all SCG and micronucleus experiments. In all cases the concurrent positive controls were found to induce a significant increase in the appropriate endpoint when compared to the concurrent sham-exposed control.

Technology	Assay	Exposure time	Result	Conclusion
ANALOG	SCG	3 hours	Positive TL only	No DNA damage
		3 hours	Negative	
		24 hours	Negative	
	MN-BN	3 hours	Negative	Genotoxic
		3 hours	Negative	
		24 hours	Positive 10 W/kg	
		24 hours	Positive 5 and 10 W/kg	
TDMA	SCG	3 hours	Equivocal TL only	No DNA damage
		24 hours	Negative	
	MN-BN	3 hours	Negative	Genotoxic
		24 hours	Positive 10 W/kg	
		24 hours	Positive 5 and 10 W/kg	
CDMA	SCG	3 hours	Equivocal TL only	Inconclusive
		24 hours	Negative	
	MN-BN	3 hours	Negative	Genotoxic
		24 hours	Positive 10 W/kg	
		24 hours	Positive 10 W/kg	
PCS	SCG	3 hours	Equivocal TL only	No DNA damage
		24 hours	Negative	
	MN-BN	3 hours	Negative	Genotoxic
		24 hours	Positive 10 W/kg	
		24 hours	Positive 10 W/kg	

Table 1. Single cell gell (SCG) and micronucleus (MN) assay results.

Results and Conclusions

The results and conclusions of all experiments are summarized in Table 1. Although equivocal or positive results were observed in some experiments using the SCG assay in no case were these results repeatable. No effect of RF fields generated by analog or digital technologies was observed following a 3-hour exposure using the micronucleus assay. A lack of genotoxic effects, measured using the SCG assay (3-

or 24-hour exposure) and the micronucleus assay (3-hour exposure) was observed for both digital and analog technologies, with and without voice modulation, and at different frequencies (837 and 1909.8 MHz). However, following a 24-hour exposure a significant increase in the number of micronucleated binucleate cells was observed for both 5 and 10 W/kg, but not at 1 W/kg. The pattern of this positive response was independent of technology or frequency.

12

SCIENTIFIC DISCUSSION OF THE RESULTS OF MICRONUCLEUS STUDIES FUNDED BY WIRELESS TECHNOLOGY RESEARCH, LLC

D.I. McRee, G.J. Hook, & R.R. Tice

Abstract

Wireless Technology Research, LLC (WTR) has funded a large number of genotoxicity studies, all of which have been negative except for studies using the micronucleus (MN) assay. In a number of experiments a significant increase in the frequency of cells with micronuclei was measured. This positive increase occurred with exposures of human blood to 837 MHz using analog, TDMA, and CDMA modulations and at 1900 MHz using a GSM modulation of 217 Hz. The increase was measured in the bottom 1 ml of media with average SARs of 10 and 5 W/kg for a duration of 24 hours. The increase was not measured at 10 and 5 W/kg for 3 hours of exposure or at 1 W/kg with 24 hours of exposure.

As part of the peer review process of the research funded by WTR, a meeting of members of the Harvard Peer Review Board and the Inter Agency Working Group was held on 31 March 1999. A number of persons were present at the meeting while others participated by teleconference. The MN exposure, dosimetry, and results were reported to the participants by WTR staff and contractors conducting the research. Much of the discussion related to the dosimetry and the distribution of SARs in the test tube and the sensitivity of MN formation to increased temperature and/or temperature gradients. From these discussions the participants recommended that certain studies be done to determine the mechanisms producing the effect and to determine the ability of other laboratories to produce similar results.

Introduction

In Wireless Technology Research, LLC (WTR) studies to determine the effects of nonionizing radiation from cellular telephones, an increase in the frequency of human blood cells with micronuclei (MN) was measured when exposed for 24 hours to relatively high specific absorption rates (SARs). For 3-hour exposures there was no increase in the frequency of MN in cells as shown in Figure 1. For 24-hour exposures, however, a significant increase in the frequency of cells with MN was measured for mean SARs of 5 and 10 W/kg but not for a mean SAR of 1 W/kg (see Figure 1).

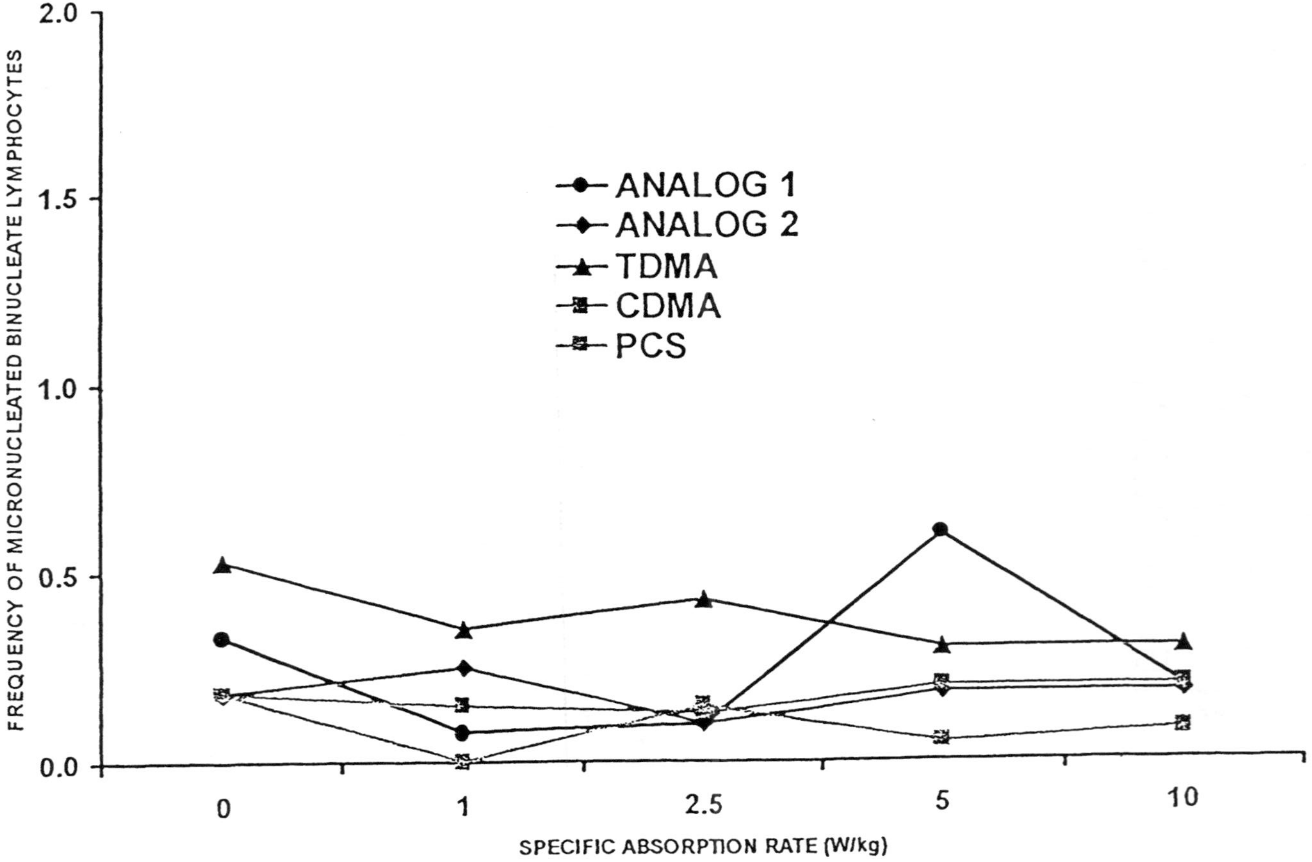

Figure 1. Frequency of micronucleus binucleate cells in human lymphocytes after 3-hour exposure.

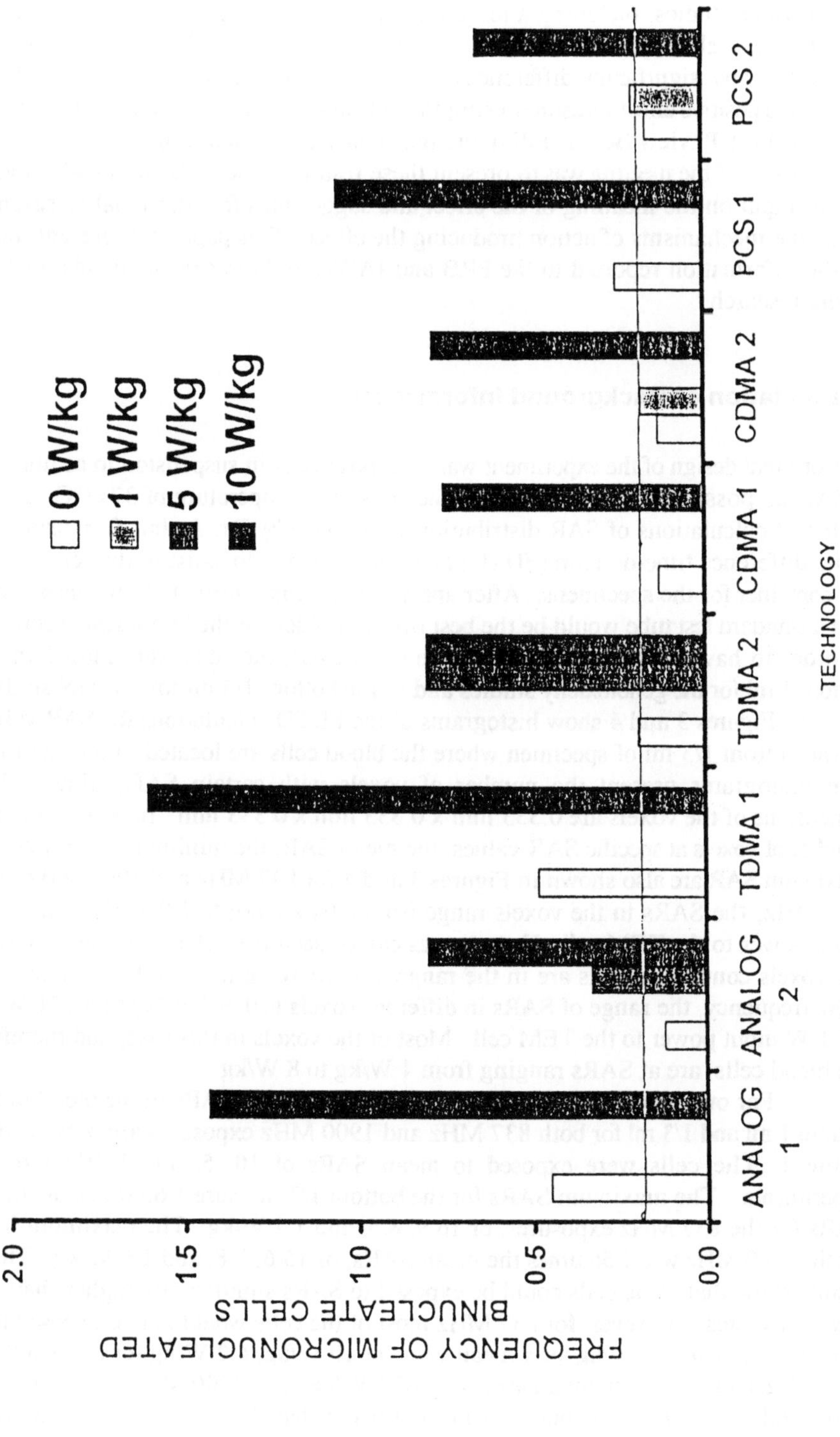

Figure 2. Frequency of micronucleate binucleate cell in human lymphocytes after 24-hour exposure.

All previous studies, including a test battery to determine genotoxicity or the single cell gel (SCG) electrophoresis assay to measure DNA migration from the nucleus, did not show any significant differences from unexposed cells. As a result of the repeatable positive effect measured using the MN assay, WTR called a meeting of the Harvard Peer Review Board (PRB) and the Inter-Agency Working Group (IAWG). The purpose of the meeting was to present these results to the PRB and IAWG and to obtain input on the meaning of the effect and suggestions for additional research to study the mechanisms of action producing the effect. This paper will present much of the information reported to the PRB and IAWG and their recommendations for future research.

Presentation of Background Information

The original design of the experiment was to expose cells in suspension to as uniform a SAR as possible while maintaining the cells at a temperature of 37±1°C. Many different calculations of SAR distribution were made by Dr. Arthur Guy using the finite-difference time-domain (FDTD) analytical method to assist in the selection of the container for the specimens. After analyzing all this information, it was decided that a standard test tube would be the best overall holder for the biological specimen. In order to have all the cells close to the same SAR, the cells were placed in the bottom 1 ml for the genotoxicity studies and in the bottom 1/3 ml for the MN studies.

Figures 3 and 4 show histograms of the FDTD calculations for SAR values in the bottom 1/3 ml of specimen where the blood cells are located in the test tube. The histograms present the number of voxels with certain SAR values. The dimensions of the voxels are 0.333 mm x 0.333 mm x 0.333 mm. In addition to the number of voxels at specific SAR values, the mean SAR, the minimum SAR, and the maximum SAR are also shown in Figures 3 and 4 for 837 MHz and 1900 MHz. For 837 MHz, the SARs in the voxels range from 0.040 W/kg to 1.036 W/kg for 1 W input power to the TEM cell. However, as can be seen from the histogram, most of the voxels containing cells are in the range of 0.40 W/kg to 0.70 W/kg. For 1900 MHz frequency, the range of SARs in different voxels is 0.301 W/kg to 8.791 W/kg for 1 W input power to the TEM cell. Most of the voxels in this case, and therefore the blood cells, are at SARs ranging from 4 W/kg to 8 W/kg.

For our experiments the mean and the maximum SARs in the blood in the bottom 1 ml and 1/3 ml for both 837 MHz and 1900 MHz exposures are presented in Table 1. The cells were exposed to mean SARs of 10, 5, and 1 W/kg in all experiments. The maximum SARs for the bottom 1/3 ml were 1.69 times the mean SARs for the 837 MHz exposures, or 16.9, 8.5, and 1.7 W/kg. The maximum SAR for the 1900 MHz was 1.56 times the mean SARs, or 15.6, 7.8, and 1.6 W/kg. These results show that some cells could be exposed to SARs significantly higher than the mean set values. However, for 837 MHz most of the cells would not be exposed to a level greater than 12.4 W/kg for a mean value of 10 W/kg, 6.2 W/kg for a mean SAR of 5 W/kg, and 1.24 W/kg for a mean value of 1 W/kg. For 1900 MHz exposure most of the cells would not be exposed to a level greater than 14.2 W/kg for a mean SAR

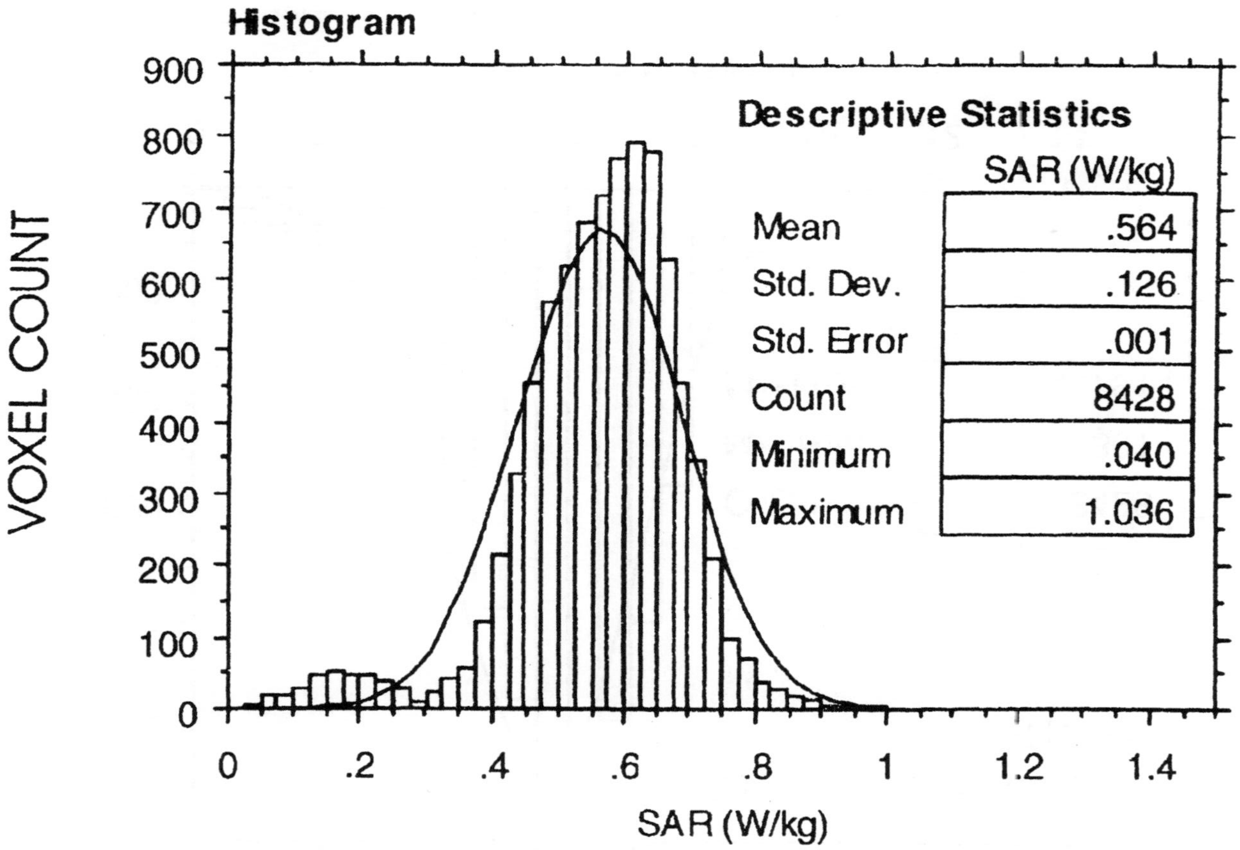

Figure 3. Histogram of calculated SAR distribution in TEM excitation plane model of suspended human blood cells in 10-ml solution with normal comparison in bottom 1/3 ml of first of a group of three tubes exposed in TEM cell at 837 MHz, 1 W input.

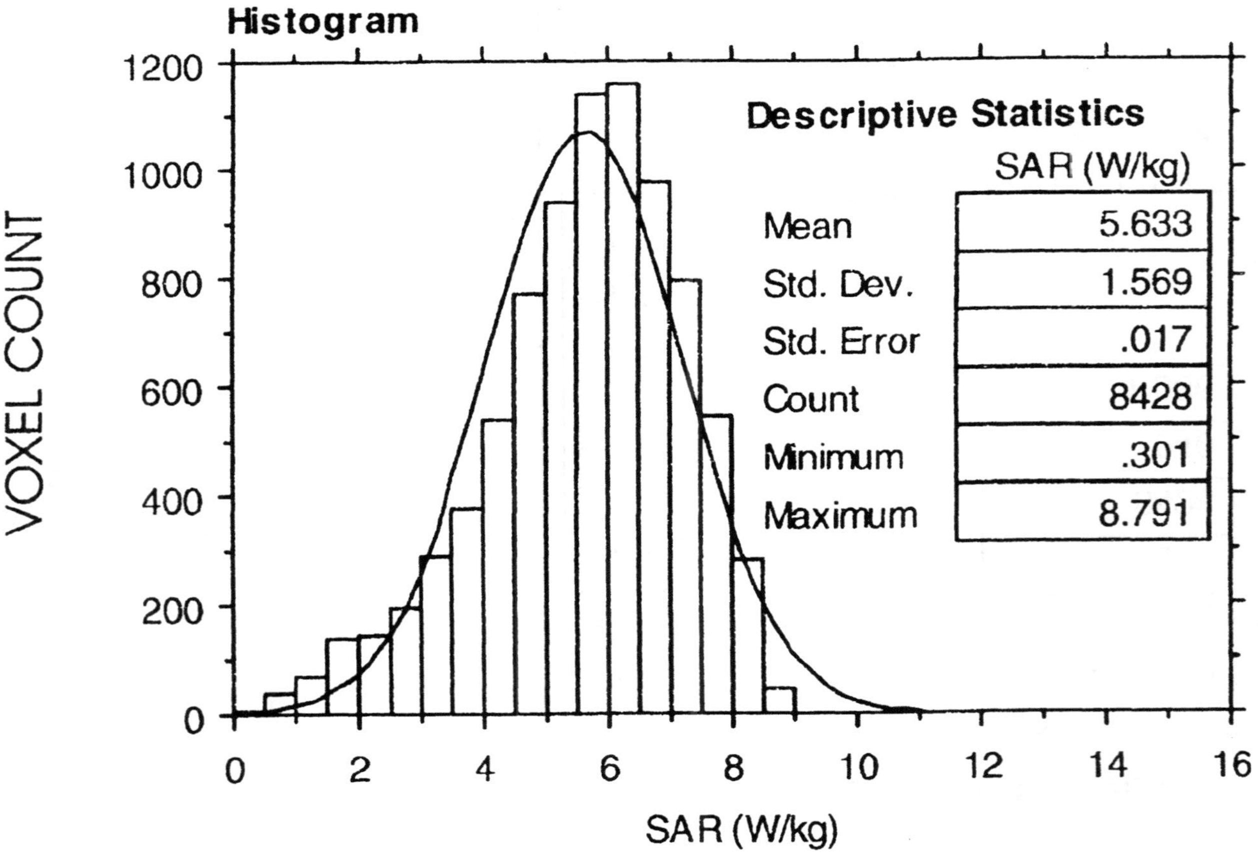

Figure 4. Histogram of calculated SAR distribution in full-size TEM full of suspended human blood cells in 10-ml solution with normal comparison in first 1/3 ml from bottom of one of a group of four tubes exposed in TEM cell at 1.9 GHz, 1 W input.

A. BOTTOM 1 ML OF BLOOD MEDIA IN TEST TUBE

837 MHz SAR (W/kg)		1900 MHz SAR (W/kg)	
Mean	Maximum	Mean	Maximum
10	26.1	10	25.9
5	13.1	5	13.0
1	2.7	1	2.6

B. BOTTOM 1/3 ML OF BLOOD MEDIA IN TEST TUBE

837 MHz SAR (W/kg)		1900 MHz SAR (W/kg)	
Mean	Maximum	Mean	Maximum
10	16.9	10	15.6
5	8.5	5	7.8
1	1.7	1	1.6

Table 1. SAR distribution in blood media.

value of 10 W/kg, 7.1 W/kg for a mean SAR value of 5 W/kg, and 1.42 W/kg for a mean SAR value of 1 W/kg.

Recommendations for Additional Research

After the above information was presented to the PRB and IAWG, several recommendations for additional research were made. The recommendations were as follows:

1. Experiments should be conducted to measure the sensitivity of MN frequency in blood lymphocytes to increased temperature without radiofrequency (RF) radiation;
2. Research projects to study the effects of 837 and 1900 MHz RF radiation on MN frequency should be done in other laboratories;
3. Efforts to evaluate the energy absorption and temperature distribution in the portion of the test tube where the blood cells are located should be continued; and
4. Additional studies should be carried out at lower SAR levels with small incremental changes in order to determine the mean and peak SAR thresholds for the occurrence of an increase in frequency of MN in blood cells.

References

Guy AW. Dosimetry studies on in vitro preparations exposed to 837 and 1900 MHz in a TEM cell. In: GL Carlo, editor. Wireless phones and health II: State of the science. Norwell, MA: Kluwer Academic Publishers; 2000.

Hook GJ. Evaluation of potential genotoxicity of radiofrequency radiation technologies using a battery of in vitro tests. In: GL Carlo, editor. Wireless phones and health II: State of the science. Norwell, MA: Kluwer Academic Publishers; 2000.

Hook GJ. Evaluation of potential genotoxicity of radiofrequency radiation technologies using thesingle cell gel electrophoresis and micronucleus assays. In: GL Carlo, editor. Wireless phones and health II: State of the science. Norwell, MA: Kluwer Academic Publishers; 2000.

IV

IN VIVO STUDIES

13

STATE OF THE SCIENCE: IN VIVO TOXICOLOGY

Transcript of a paper presented by M. Repacholi

Edited by WTR Staff

Animal studies are probably the most useful type of research that can be applied to health risk assessments because of the precision that can be incorporated into these studies. The World Health Organization (WHO) has conducted a number of electromagnetic field (EMF) reviews to cover all of the areas that need to be covered. A radiofrequency (RF) meeting in Munich on low-level fields occurred in November 1996. Its review has already been published in *Bioelectromagnetics*. The Static and ELF meeting in Bologna is another review which was also published in *Bioelectromagnetics* in 1999. We have already had meetings in Eastern Europe to try and capture the Russian literature which is out there but has been largely ignored in the health risk assessment process. Due to these meetings, we have proceedings that are coming out in both Russian and English, as well as a summary report which will be published over the next 6 months. We have also had a meeting in Austria on psychosocial effects. The proceedings for that meeting have already been published, and a summary report is currently being prepared. We met with the Chinese in Beijing last month to get an overview of what literature is available there and subsequent reports will be coming out soon. A meeting was held on the intermediate frequency range between the low frequencies and the radiofrequencies, that is, the frequency range of 300 Hz to about 10 MHz. There will be a proceedings and summary report published from that. The next major meeting of concern for this group will be our last review of the literature on pulsed, low-level radiofrequency fields in Sicily in November of this year. It will be a major review and update of our literature on RF fields.

We do know that effects occur from specific absorption rates (SARs) above 4 W/kg. This seems to be the threshold for thermally induced effects. There are a number of effects still not established such as memory loss, cancer changes, blood pressure, subjective effects, and blood-brain barrier effects. These are areas on which we need to concentrate.

I have only looked at studies after 1980 because prior to 1980, and probably prior to 1990, the dosimetry of many of these reports was poor, while others had methodological flaws. There have been few animals that could successfully be used to test for positive or negative effects. The criteria for assessing these animal studies are those used by the International Agency for Research on Cancer (IARC). Some studies reporting a negative effect had flaws in the methodology. For example, it may have been that the study could only produce a negative effect because it did not have the sensitivity within the study to produce a positive effect. In this case, that study is not useful for health risk assessment. High-quality studies producing both positive and negative effects must be performed. There are a couple of good studies by Saunders et al. looking at dominant lethal assays and other mutational studies that give some confidence that there does not seem to be an initiation effect in the process of carcinogenesis.

If we look at the chromosome aberration studies, we again see that there is a mixed bag of studies. Some indicate increases in chromosomal aberrations, but some of these studies are relatively poor, mainly because of the dosimetry. It is not difficult to do chromosome aberration studies, but the dosimetry can be such that you can get such non-uniformity within the exposed system that you are not sure to what you are exposing. By and large, though, you generally see that there are no chromosomal aberrations resulting from the majority of studies. There are, however, some that indicate an increase in chromosomal aberrations in occupationally exposed human subjects. Another occupational study by Garson et al. in Australia looking at human lymphocytes of linemen found no increase in chromosome aberrations. It would seem that given the basic physics of RF fields, without sufficient heating, that you would not expect chromosomal aberrations to occur.

We have seen some increase in micronuclei in mouse hepatocytes, but again these are somewhat older studies and we tend to look towards more recent studies to see if there are going to be micronuclei. This is a very recent study in which they are looking at ultra-wideband radiation, 15-minute exposure to whole-body SARs of 37 mW/kg and they did not find any micronuclei. This is really quite a good study with good dosimetry. If we have a look at sister chromatid exchanges, with the exception of Evert Cheska's paper, no effect is being found on sister chromatid exchanges. If we have a look at DNA damage and repair studies, most of the studies are indicating no effect on DNA breaks or repair. When you look at a weight-of-evidence approach to these studies, you get a fairly good idea that maybe there is not anything occurring. If we look at cancer initiation studies, four of these studies have been done and, in general, you see that there have not been any effects, but there have been increases in this transgenic model. For example, the study by Chou was a 25-month study showed an increase in the incidence of primary malignancies, but this was an internal inconsistency of this study. The most recent promotional studies have not found any effect on cancer promotion when you add an initiating agent. Finally, within the longevity studies, there was an early study indicating that you get an increase in life span. In general, however, there is no effect on life span. Certainly one of the better studies is by Chou and the 25-month study which indicated no ill effect.

The study that was carried out in Adelaide used groups of 100 transgenic mice having extra *pim*-1 genes, so that the difference was a genome. This number was selected because it was determined in power calculations to allow the detection of the doubling of the incidence of cancer at the p=.05 level. It was a specific pathogen-free facility and double-blind design from the onset. The results were that non-lymphoblastic lymphomas increased quite dramatically over the control levels. The lymphoblastic lymphomas also increased, but the numbers were too few to get a significant result. Overall, we have a tremendous increase in predominately B-cell lymphomas that occurred and there have been a lot of extrapolations of these results. It is imperative that study results are independently replicated before they can be established. Replication and extension is needed so you can see if there is a dose-response, requiring the dosimetry to be more responsive. We also need to know what it means when you get a result in transgenic animals and how it applies to humans. This study could be designed as an assay technique useful for determining mechanisms involved.

A team of scientists in Australia have been trying to replicate and extend this particular study. There has been a replication in Europe which is currently being reviewed by the European Commission. Again, the study needs to be double-blind but it should be extended to 24 months from the 18 months using at least at three different RF exposures and perhaps transgenic mice models. The models that could be used are the p53 as a well characterized mouse model.

In conclusion, the Royal Society of Canada concluded in their literature review that the vast majority of studies on genotoxic effects of RF have proven to be negative, suggesting that there is no reason for concern on the part of the general public. The review also indicated that there was a need for further research which is quite reasonable under the circumstances. The WHO's response to the Canadian review was that although hazards from exposure to high-level thermal effects are established, no known health hazards are associated with exposure to RF sources emitting fields too low to cause significant temperature rise. However, there are biological effects from these low levels which have been identified and which need replication and further study.

14

THE CARCINOGENIC POTENTIAL OF 835.62 MHZ FMCW AND 847.74 MHZ CDMA RADIATIONS: ONGOING STUDIES

J.L. Roti Roti, R.S. Malyapa, R. Higashikubo, P. Zhang, L. Li, W.L. Straube, E.G. Moros, & W.F. Pickard

The number of mobile phone users is rapidly increasing, with the number expected to exceed 200 million worldwide by the year 2000. Because of the large number of individuals exposed to the radiofrequency (RF) signals from cellular phones, the concern for any detrimental effects from these exposures is magnified. This concern would be magnified further if any such effects could conceivably contribute to a carcinogenic process. Conversely, the availability of convincing data demonstrating the lack of such effects would be assuring to both industry and the public that these radiations have minimal risk in this regard. As reviewed at the previous Wireless Technology Research, LLC (WTR) Colloquium, the available scientific literature failed to support the former, but was not sufficient to support the latter (Carlo 1998).

For the past 5 years we have been engaged in an ongoing study to assess the biological effects of RF radiations relevant to mobile communication with emphasis on those effects which could be related to the induction, promotion, or progression of cancer. To accomplish this goal we have undertaken a series of in vivo and in vitro projects. These projects address various aspects of the operationally defined cancer paradigm. The projects include a study of implanted brain tumor growth (in vivo), direct measures of initiation and promotion, DNA damage, and proto-oncogene expression. The choice of projects and the endpoints to be measured were based on the operational model of carcinogenesis with emphasis on possible epigenetic carcinogenic effects. The operational steps of the carcinogenic process are initiation, promotion, and progression. Presumably if RF radiation could contribute to the carcinogenic process it would have to positively affect one or more of these steps. Most known carcinogens contribute to the initiation step and are genotoxic. Therefore, a major effort has been made by us and other groups to determine if there are genotoxic effects of nonthermal RF exposures, particularly DNA damage. There are relatively few known tumor promoters, TPA being one. However, most of the proposed, but untested, mechanisms by which RF exposure can contribute to the carcinogenic process are epigenetic or promotional effects. Therefore, we expended significant work in this regard; this work is of necessary research effort, because there are no general epigenetic carcinogen tests. While there are known mechanisms for tumor progression, there are few, if any, specific carcinogens that target this step. Therefore, we addressed the possibility of tumor bed effects and measured the growth of implanted 9L tumor cells in the brains of rats chronically exposed to RF radiation. By means of this comprehensive approach we hope to minimize the chances of missing a critical biological effect of RF radiation.

To conduct the studies described in this chapter we (WFP, EM, & WS) designed and built two exposure systems (Moros et al. 1998, 1999a, 1999b; Pickard et al. 1999). Since the description and characterization of these systems have been published, only a few key features will be mentioned here. Both systems were designed to expose large numbers of specimens, rats for the in vivo irradiators and T-75 culture flasks for the in vitro irradiators. The in vitro exposure system was based on Radial Transmission Line (RTL) irradiators, which consisted of two parallel aluminum plates. A conical antenna occupied the center of the plate with 16 T-75 flasks surrounding the antenna in a circular array. The irradiated cells were in a monolayer parallel to the bottom plate, separated from it by an air layer, the plastic bottom of the tissue culture flask, and an aluminum plate to provide additional dielectric material. Temperature monitoring of the culture flasks was continuous throughout the exposure period for time intervals from less than 1 hour to 6 weeks. The temperature was controlled at 37.0± 0.3°C. Without the aluminum plate the specific absorption rate (SAR) was 0.6 W/Kg, while insertion of the plate increased the SAR to 2.4 W/Kg. The total capacity of the system was such that we could expose 32 flasks to each of two signals and a sham simultaneously. The in vivo exposures were conducted in 12 chamberettes, which held 40 rats each. The rats were placed in restraint devices in a circular array about a central antenna with the rats' heads oriented toward it. The mean SAR for this exposure system was 0.7 W/Kg to the brain. These exposure systems proved to be mechanically and thermally reliable and relatively easy to use by laboratory personnel.

Genotoxicity Studies

For the purposes of this review results from two types of genotoxicity studies will be described. First, measurements of DNA damage induction tissue culture cells and in rat brain cells exposed in vivo will be described. Second, we will consider measurements of in vitro transformation as an initiator. Both of these are relevant to the initiation step of the transformation process, because the latter is a direct measure of it, and induction and repair of DNA damage is the initiation mechanism most often observed.

Concern that RF fields might have a genotoxic component was heightened by the publications of Lai and Singh (Lai & Singh 1995, 1996). These authors reported that DNA from rat brain cells exposed to 2450 MHz CW and PW at a SAR of 1.2 W/Kg showed more migration in single cell gel (SCG) electrophoresis than that measured in brain cells from the sham-exposed rats. These results were interpreted by Lai and

Singh to imply that microwave irradiation could induce DNA single- and double-strand breaks (Lai & Singh 1996). However, no definitive mechanism for the induction of the putative DNA damage was proposed, nor was any attempt made to corroborate its presence by an alternative method. Therefore, attempts to confirm that such damage was induced by RF exposure utilized the same or similar methods to detect it.

We began by using both in vivo and in vitro approaches so that if any DNA damage was detected we could immediately begin to determine the mechanism of its induction. For the in vivo approach we exposed an immortalized, non-transformed cell line (C3H 10T-1/2, ATCC) and a human brain tumor cell line (U87, ATCC). These cells were exposed to 2450 MHz CW at a SAR of up to 1.9 W/Kg (Malaypa et al. 1997a) and to 835.62 MHz FMCW and 847.74 MHz CDMA (Malaypa et al. 1997b) at a SAR of 0.6 W/Kg for 2, 4, and 24 hours. The C3H 10T-1/2 cells were exposed in both the exponential and plateau growth states. The principle positive control was ionizing radiation (γ- or X- rays). We were able to detect DNA damage following γ-ray doses as low as 0.6 cGY, which was comparable to that published by Singh et al. (1994, 1995; Malyapa et al. 1998a). Another positive control was UV laser light on HO33342 stained cells (Roti Roti et al. 1998). We did not detect any DNA damage following exposure to any of these RF radiations following 2, 4, or 24 hours or 2 hours with a 4-hour further incubation. In order to sensitize cells to the induction of DNA damage we labeled them with 4 μM BrdU prior to RF exposure. Fluorescent light on cells that had incorporated BrdU produced measurable DNA damage, but exposure of these cells to either of the above RF radiations resulted in no detectable DNA damage. While these studies suggested that these RF radiations did not cause DNA damage to exposed cells, the issues of the low SARs used for the 835.62 MHz FMCW and 847.74 MHz CDMA radiations, the detection of DNA protein cross-links, and the effects of 2450 MHz in vivo remained unresolved.

To determine if 2450 MHz CW microwaves caused DNA damage in the brain cells of exposed rats we placed the rats in two (one being a sham) Guy-type irradiators for 2 hours (as described in Malyapa et al. 1998b). Rats were either euthanized immediately after exposure or kept alive for 4 hours after exposure. DNA damage was measured by the SCG assay in excised hippocampal and cortex cells. Throughout these studies the only DNA damage that could be detected was due to the method of euthanasia used prior to brain dissection. To resolve possible differences in methodology we, in collaboration with Dr. Bernard Veyret's group, conducted the next study using two versions of the SCG assay in a side-by-side double-blind comparison. In this study rats were exposed to 2450 MHz PW for 2 hours. In addition, one group of rats was irradiated with 1 Gy of γ-rays and brain cells excised immediately after exposure. Half of the cells were processed following the method of Singh (Singh et al. 1994, 1995) and the other half by the Olive method (Olive et al. 1992). The results (LaGroye et al. 1999) showed that while both methods could detect DNA damage following the 1 Gy radiation dose, neither detected DNA damage following exposure to 2450 MHz PW.

Further in vitro studies were designed to determine if 2450 MHz CW can induce DNA cross-linking and if 835.62 MHz FMCW and 847.74 MHz CDMA at SARs of 3 to 5 W/Kg can induce alkali labile DNA damage. The former study by Lagroye et

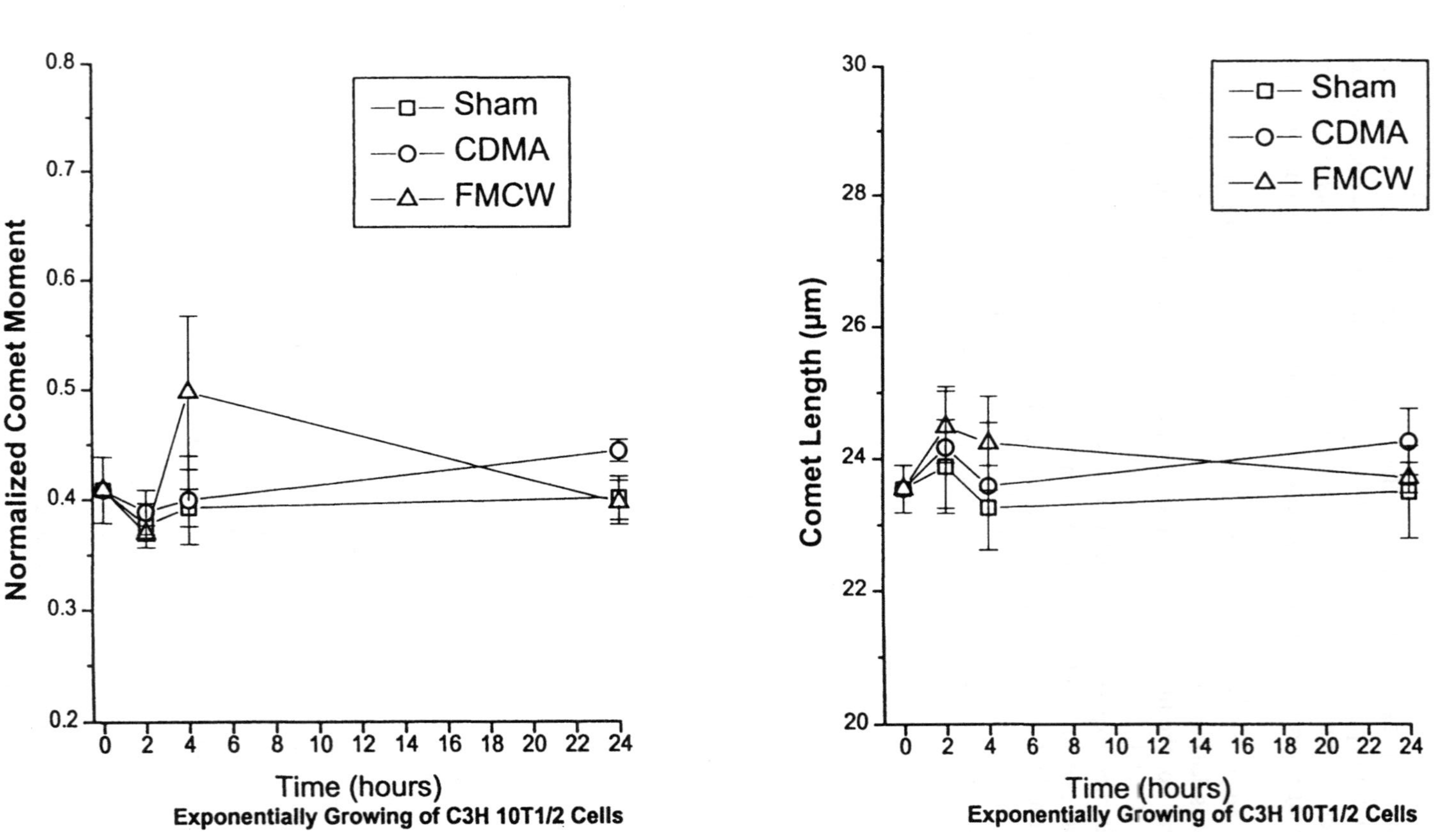

Figure 1. Effect of 847.74 MHz CDMA and 835.62 MHz FMCW at a minimum SAR 3.2 W/Kg on induction of DNA damage in exponentially growing C3H 10T-1/2 cells detected by the alkaline comet assay. Cells were exposed for 2, 4, or 24 hours at 37±0.3°C. Data points represent the average means from three independent experiments. The error bars represent ± SEM.

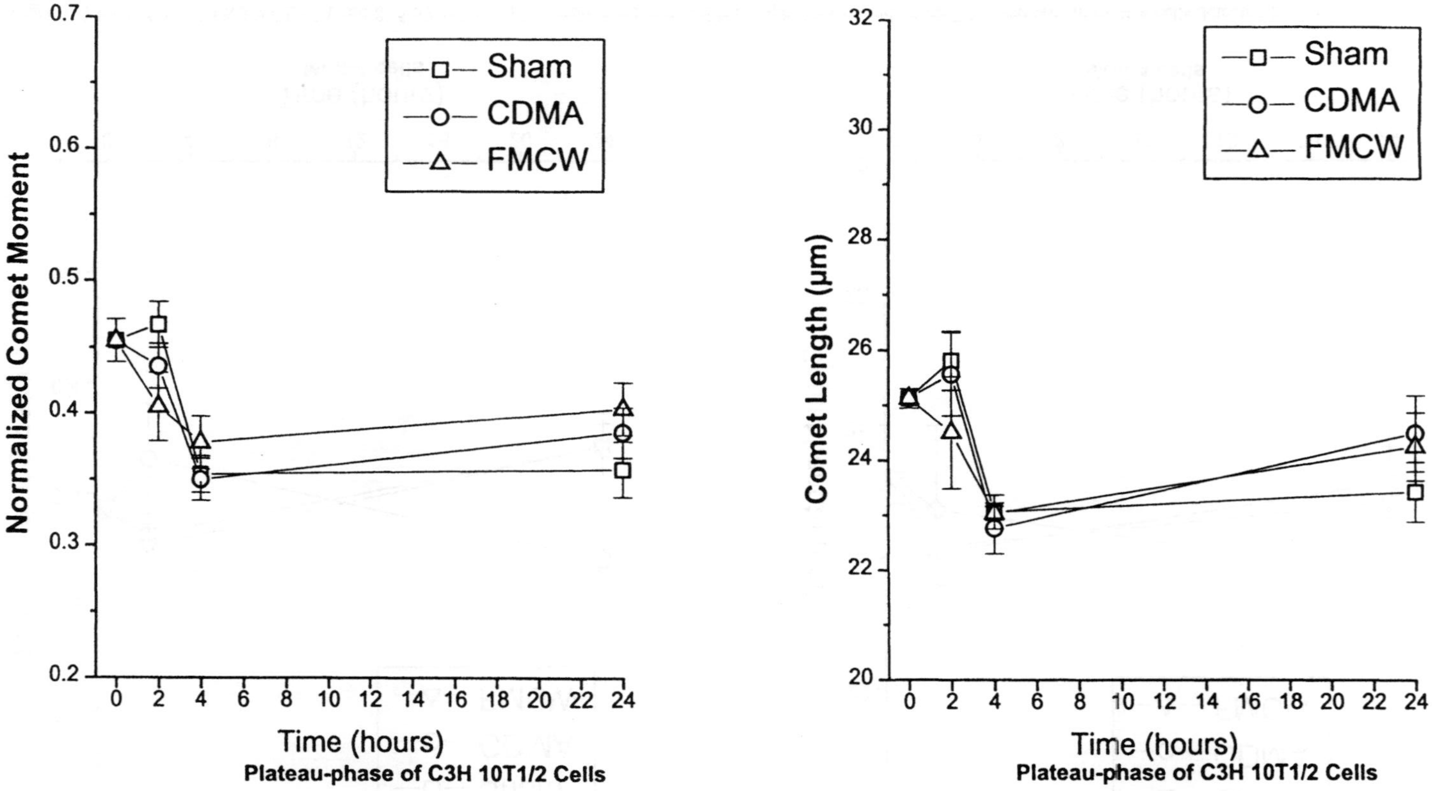

Figure 2. Effect of 847.74 MHz CDMA and 835.62 MHz FMCW at a minimum SAR 3.2 W/Kg on the induction of DNA damage in plateau-phase C3H 10T-1/2 cells detected by the alkaline comet assay. Cells were exposed for 2, 4, or 24 hours at 37±0.3°C. Data points represent the average of means from three independent experiments. The error bars represent ± SEM.

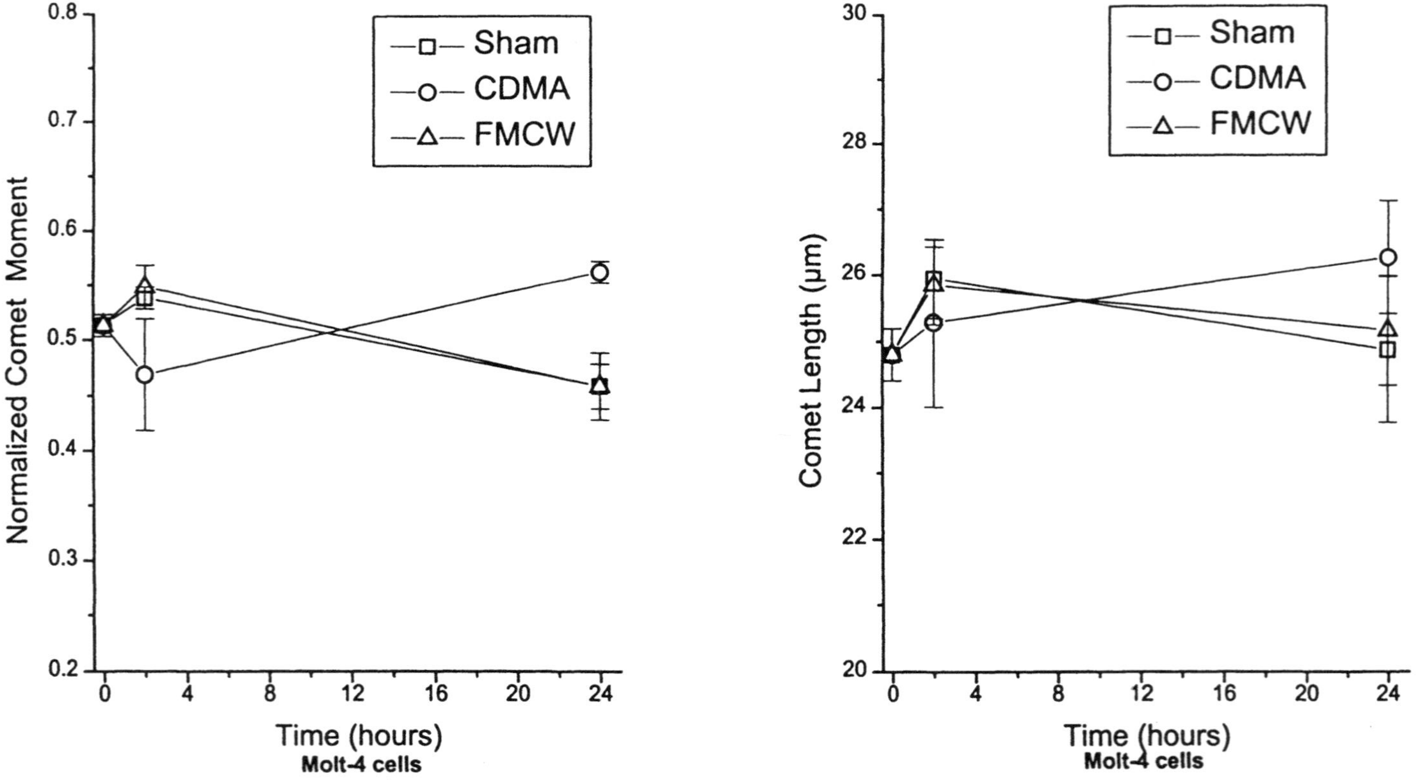

Figure 3. Effect of 847.74 MHz CDMA and 835.62 MHz FMCW radiofrequency fields at high SAR on induction DNA damage in molt-4 cells detected by the alkaline comet assay. Cells were exposed for 2 and 24 hours at 37°C. Data points represent the average of means from three independent experiments. The error bars represent ±1 SEM.

al. (see Chapter 10 in this volume) showed that no DNA cross-links were detected following exposure to 2450 MHz CW. For the latter study we exposed C3H 10T-1/2 cells and molt-4 cells to these signals for 2 and 24 hours. The results (Figures 1, 2, & 3) show that after either 2 or 24 hours of exposure, both the comet length and comet moment were essentially the same as sham for both cell lines. Thus, higher SAR exposures did not induce DNA damage. Together these results demonstrate that a broad effort failed to detect alkaline labile DNA damage following RF exposure. Thus, the weight of new results supports the conclusion that these radiations do not cause DNA damage.

The in vitro transformation assay, based on the C3H 10T-1/2 system was used to determine if chronic exposure to 835.62 MHz FMCW or 847.74 CDMA at 0.6 W/Kg initiates neoplastic transformation. Approximately 300 clonogenic C3H 10T-1/2 cells were seeded into 32 T75 flasks for each exposure condition, 24 hours before the beginning of the exposure period. Both sham-exposed and incubator controls were used to determine if placing cells in the RTLs had any effects. The preliminary experiments for this study were presented at the previous WTR Colloquium (Roti Roti et al. 1998). These studies showed the expected ionizing radiation dose-response and response to 0.1 μg/ml TPA (Roti Roti et al. 1998). There was no difference in the measured transformation frequencies between incubator controls and sham exposed for any experimental condition, showing that these two conditions were equivalent in terms of the induction of neoplastic transformation. After a 1-week exposure to 835.62 MHz FMCW, 847.74 MHz CDMA, or the sham RTL, the flasks were maintained in the incubator for another 6 weeks to express transformed foci. Foci were classified into types I, II, & III according to established procedures (International Agency for Research on Cancer/National Cancer Institute/Environmental Protection Agency Working Group 1985) with types II & III being scored as neoplastic, based on their reported ability to form tumors in nude mice (Hall and Miller 1977). Three of these experiments were completed. We then conducted four more experiments in which we added 0.1 mM of the known promoter, TPA, to the flasks after exposure to enhance the expression of any cells that may have been initiated. However, for all of these conditions the measured transformation frequency was not significantly different from the background, suggesting that no initiation was occurring. We therefore pooled the results of these seven experiments and obtained the following transformation frequencies: sham, 4.60E-5 + 2.00E-5; FMCW, 6.09E-5 + -2.36E-5; and CDMA, 5.40E-5 + 2.42E-5. Thus, neither 835.62 MHz FMCW nor 847.74 MHz CDMA at a SAR of 0.6 W/Kg were able to induce measurable levels of neoplastic transformation.

As an overall conclusion of our studies thus far, we have failed to detect any genotoxic effects attributable to these RF radiations. Our results, along with that of others presented at the Colloquium and reviewed at the previous Colloquium, suggest that RF radiations are not directly genotoxic and, therefore, not likely to be initiators of the neoplastic process. However, the question of micronucleus formation discussed at this meeting clearly remains to be resolved.

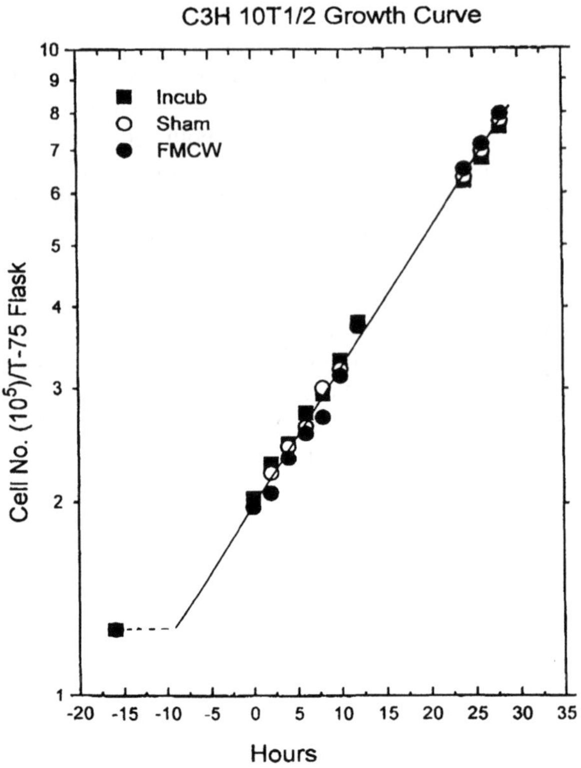

Figure 4. The lack of effect of 835.62 MHz FMCW on cell proliferation. C3H 10T-1/2 cells were subcultured in T75 flasks 16 hours before they were placed in the RTLs. One third of the flasks were placed in sham RTLs, 1/3 in RTLs radiation 835.62 MHz FMCW at 0.6 W/Kg, and 1/3 left in the incubator. Two flasks were counted for cell numbers 2 hours for the first day and part of the second day. The averages of the cell numbers from the two flasks are plotted (ordinate) as a function of time ordinate in the RTLs. While typical lag in the onset of proliferation is seen prior to cell proliferation, no effect of the FMCW radiation can be seen.

Promotion Studies

For the purpose of this report we will discuss the results of studies of cell proliferation and proto-oncogene expression. In addition we will present the results of direct measures of promotion of neoplastic transformation.

The rate of proliferation of C3H 10T-1/2 cells was unchanged when cell cultures were exposed to 847.74 MHz CDMA at a SAR of 0.6 W/Kg (Figure 4). The same results were obtained with 835.62 MHz FMCW (data not shown). Studies with HeLa cells showed no changes in the rate of cell proliferation, the rate of DNA synthesis, the rate of RNA synthesis, or the rate of protein synthesis when cell cultures were exposed to either 835.62 MHz FMCW or 847.74 MHz CDMA at a SAR of 0.6 W/Kg

(Mackey, unpublished results). Further studies with C3H 10T-1/2 cells showed that the rate of cell-cycle progression as measured by five parameters was unchanged when cell cultures were exposed to either 835.62 MHz FMCW or 847.74 MHz CDMA at a SAR of 0.6 W/Kg (Higashikubo, unpublished results). These results suggest that these RF radiations do not directly affect cell proliferation.

To determine if exposure to either 835.62 MHz FMCW or 847.74 MHz CDMA at a SAR of 0.6 W/Kg induced a stress response, we measured the expression of the early response genes (c-*fos*, c-*jun*, and c-*myc*) and the DNA binding activity of the transcription factors (AP-1, AP-2, and NFkB). Because early concerns were for chronic exposures, we studied the effects of exposures ranging from 24 hours to 1 week. We measured the effects of 835.62 MHz FMCW and 847.74 MHz CDMA on these parameters as cells entered the plateau phase from exponential growth and as cells entered the cell cycle from quiescence. The results (Goswami et al. 1999) showed that neither 835.62 MHz FMCW nor 847.74 MHz CDMA at a SAR of 0.6 W/Kg had any significant effect on proto-oncogene expression during transit from quiescence to the proliferation. Also, these exposures did not affect the binding activity of AP-1, AP-2, and NFkB during this transition. Similarly, 835.62 MHz FMCW and 847.74 MHz CDMA did not affect the binding of these transcription factors during the transition from proliferation to plateau. While the expression of c-*jun* and c-*myc* were not significantly affected during this transition, the levels of c-*fos* mRNA were 2-fold higher in the cells exposed to 835.62 MHz FMCW and 1.45-fold higher in the cells exposed to 847.74 MHz CDMA than in the sham exposed. Since c-*fos* expression is upstream from c-*jun* and c-*myc*, it appears that these RF signals did not induce a general stress response. However, it may be that expression of specific gene(s) could be affected by these RF exposures. Further, studies of the effects of acute exposures (i.e., 20-120 minutes) at higher SARs remain to be completed.

The in vitro transformation assay, based on the C3H 10T-1/2 system, was used to determine if chronic exposure to 835.62 MHz FMCW or 847.74 MHz CDMA at 0.6 W/Kg could promote neoplastic transformation. The protocols for these studies were similar to those described above for the initiation experiments, except that the flasks were exposed to the RF signals throughout the 6-week expression period. To initiate a reasonable number of transformed foci, C3H 10T-1/2 cells were initiated with 4.5 Gy of γ-rays, which induced a 15-fold increase above background transformation frequency, independent of whether the cells were in sham RTLs or in the incubator. Data were pooled from six independent experiments. The overall transformation frequencies were 7.35E-4 + 0.90E-4 following 4.5 Gy, 6.64E-4 + 0.92E-4 following 4.5 Gy and exposed to 0.6 W/Kg of 835.62 MHz FMCW radiation for 6 weeks, and 6.02E-4 + 0.85E-4 following 4.5 Gy and exposed to 0.6 W/Kg of 847.74 MHz CDMA for 6 weeks. Thus, neither 835.62 MHz FMCW nor 847.74 MHz CDMA at 0.6 W/Kg had any significant effect on promotion of neoplastic transformation in the C3H 10T-1/2 system.

Overall, the work completed thus far does not support the possibility that RF radiation from mobile phones has any epigenetic effects related to neoplastic transformation. However, any conclusion on this matter must wait for completion of the ongoing work in ours and several other laboratories because particular epigenetic mechanisms are uniquely defined. Therefore, more data on any robust effects of RF radiations on cells will be critical in this regard.

Effects on Tumor Growth

We used the transplantable 9L rat brain tumor system to determine if chronic exposure to 835.62 MHz FMCW or 847.74 MHz CDMA at a slot-average SAR of 0.75 W/Kg in the brain could have either an effect on the ability of tumor cells to establish a tumor or the time required for the tumor to reach a critical size. A custom-designed animal irradiation system was built so that 480 rats could be simultaneously exposed to 835.62 MHz FMCW, 847.74 MHz CDMA, or sham with 160 animals in each exposure group. The exposure chamberettes have been described previously (Moros et al. 1998, 1999a). In order to test for any possible effects of exposure on tumor establishment and subsequent growth, rats were exposed for 4 weeks prior to and up to 150 days after tumor cell implantation. Between 1 and 100 clonogenic 9L tumor cells, as measured by in vitro colony forming ability, were implanted into the brains of Fisher 344 rats. The results (Higashikubo et al., in press) showed that when the fraction of rats developing tumors was plotted as a function of the number of tumor cells injected, there was no difference between the sham-exposed rats and those exposed to either 835.62 MHz FMCW or 847.74 MHz CDMA. Based on the survival curves of the sham-exposed animals, it was apparent that these should be grouped according to the number of clonogenic cells injected as follows: group I, 2 to 10 cells injected; group II, 11 to 36 cells injected; and group III, 37 to 100 cells injected. For the sham-irradiated animals, the median survival times and final surviving fractions were 70 days and 27% for group I, 52 days and 14% for group II, and 45 days and 0% for group III. For the animals exposed to 835.62 MHz FMCW, the median survival times and final surviving fractions were 58 days and 18% for group I, 53 days and 16% for group II, and 43 days and 10% for group III. For the animals exposed to 847.74 MHz CDMA, the median survival times and final surviving fractions were 71 days and 26% for group I, 53 days and 14% for group II, and 41 days and 5% for group III. Analysis of these survival curves showed no statistically significant differences between sham-exposed animals and those exposed to either 835.62 MHz FMCW or 847.74 MHz CDMA (Higashikubo et al., in press). Thus, these RF radiations had no effect on the ability of tumor cells to establish a tumor or on their subsequent growth in situ.

Conclusions

Any overall conclusions from this work can only be based on the completed portion and, therefore, must be qualified as such. However, extensive work in specific areas warrants conclusions regarding the role of the studied RF signals in certain aspects of the neoplastic process. The evidence is becoming increasingly convincing that 835.62 MHz FMCW and 847.74 MHz CDMA at 0.6 W/Kg are not genotoxic and are probably not initiators of the neoplastic process. The studies completed so far show no evidence for an epigenetic contribution from either 835.62 MHz FMCW or 847.74 MHz CDMA at 0.6 W/Kg to the neoplastic process or to an altered tumor growth rate. However, ongoing work needs to be completed and certainly more work will need to be undertaken before any possible epigenetic mechanisms associated with RF exposures can be excluded.

Acknowledgment

This work was sponsored by a contract from the Motorola Corporation. The authors would like to thank Ms. Kathy Bles for assistance with word processing.

References

Carlo GL, editor. Wireless phones and health: Scientific progress. Norwell, MA: Kluwer Academic Publishers; 1998.

Goswami PC, Albee LD, Parsian AJ, Batey JD, Moros EG, Pickard WF, Roti Roti JL, & Hunt CR. Protooncogene mRNA levels and activities of multiple transcription factors in C3H 10T1/2 murine embryonic fibroblasts exposed to 835.62 and 847.74 MHz cellular phone communication frequency radiation. Radiat Res 1999; 151: 300-309.

Hall EJ & Miller RC. The how and why of in vitro oncogenic transformation. Radiat Res 1977; 87: 452.

Higashikubo R, Culbreth VO, Spitz DR, LaRegina MC, Pickard WF, Straube WL, Moros EG, & Roti Roti JL. Radiofrequency electromagnetic fields have no effect on the in vivo proliferation of 9L brain tumors. In press.

International Agency for Research on Cancer/National Cancer Institute/Environmental Protection Agency Working Group. Cellular and molecular mechanisms of cell transformation and standardization of transformation assays of established cell lines for the prediction of carcinogenic chemicals: Overview and recommended protocols. Cancer Res 1985; 45: 2395-2399.

Lai H & Singh NP. Acute low intensity microwave exposure increases DNA single-strand breaks in rat brain cells. Bioelectromagnetics 1995; 16: 207-210.

Lai H & Singh NP. Single- and double-strand DNA breaks in rat brain cells after acute exposure to radiofrequency electromagnetic radiation. Int J Radiat Biol 1996; 69: 513-521.

Lagroye I, Wettring BA, Moros EG, Straube WL, Pickard WF, & Roti Roti JL. Measurement of alkali labile DNA damage and cross-links following 2450 MHz microwave and low dose gamma irradiation in vitro. In: GL Carlo, editor. Wireless phones & health II: State of the science. Norwell, MA: Kluwer Academic Publishers; 2000.

Malyapa RS, Ahern EW, Straube WL, Moros EG, Pickard WF, & Roti Roti JL. Measurement of DNA damage after exposure to 2450 MHz electromagnetic radiation. Radiat Res 1997a;

148: 608-617.

Malyapa RS, Ahern EW, Straube WL, Moros EG, Pickard WF, & Roti Roti JL. Measurement of DNA damage after exposure to electromagnetic radiation in the cellular communication frequency band (835.62 and 847.74 MHz). Radiat Res 1997b; 148: 618-627.

Malyapa RS, Bi C, Ahern EW, & Roti Roti JL. Detection of DNA damage by the alkaline comet assay after exposure to low-dose gamma radiation. Radiat Res 1998a; 149: 396-400.

Malyapa RS, Ahern EW, Bi C, Straube WL, LaRegina M, Pickard WF, & Roti Roti JL. DNA damage in rat brain cells after in vivo exposure to 2450 MHz electromagnetic radiation and various methods of euthanasia. Radiat Res 1998b; 149: 637-645.

Moros EG, Straube WL, & Pickard WF. A compact shielded exposure system for the simultaneous long-term UHF irradiation of forty small mammals: I. Electromagnetic and environmental design. Bioelectromagnetics 1998; 19: 459-468.

Moros EG, Straube WL, & Pickard WF. Compact shielded exposure system for the simultaneous long-term UHF irradiation of forty small mammals: II: Dosimetry. Bioelectromagnetics 1999a; 20: 81-83.

Moros EG, Straube WL, & Pickard WF. The radial transmission line as a broad-band shielded exposure system for microwave irradiation of large numbers of culture flasks. Bioelectromagnetics 1999b; 20: 65-80.

Olive PL, Wlodek D, Durand RE, & Banath JP. Factors influencing DNA migration from individual cells subjected to gel electrophoresis. Exp Cell Res 1992; 198: 259-267.

Pickard WF, Straube WL, Moros EG, & Fan X. Simplified model and measurement of specific absorption rate distribution in a culture flask within a transverse electromagnetic mode exposure system. Bioelectromagnetics 1999; 20: 183-193.

Roti Roti JL, Malyapa RS, & Ahern EW. Considerations regarding the use of in vitro transformation and alkaline comet assays to assess the carcinogenic potential of RFR. In: GL Carlo, editor. Wireless phones and health: Scientific progress. Norwell, MA: Kluwer Academic Publishers; 1998. pp. 113-126.

Singh NP, Stephens RE, & Schneider EL. Modifications of alkaline microgel electrophoresis for sensitive detection of DNA damage. Int J Radiat Biol 1994; 66: 23-28.

Singh NP, Graham MM, Singh V, & Khan A. Induction of DNA single-strand breaks in human lymphocytes by low doses of γ-rays. Int J Radiat Biol 1995; 68: 563-569.

15

EFFECTS OF GSM-900 MICROWAVES ON RAT MODELS

Transcript of a paper presented by R. Anane

Edited by WTR Staff

In the context of the research on health effects of mobile communications, we in the laboratory at the University of Bordeaux are investigating the effects of exposure to 900 MHz GSM microwaves on different rat models. Before the start of the biological studies, we developed two exposure systems. The first was a near-field loop antenna operating at 900 MHz for localized irradiation of rat heads. The second system was a far-field base station antenna operating at 900 MHz within two 3 m x 3 m x 3 m anechoic chambers. Finite-difference time-domain (FDTD) calculations were made to ensure specific absorption rate (SAR) distribution uniformity.

We conducted our first experiment in co-promoting dimethylbenz[a]anthracene (DMBA) tumors. We used 128 female Sprague Dawley rats from 50-60 days old. There were seven different groups: a control cage, a sham-exposed group, and five groups exposed to 0.1, 0.7, 1.4, 2.2, and 3.5 W/kg; the whole-body exposure was for 2 hours per day 5 days per week for 9 weeks. We determined different parameters such as body weight, tumor volume, tumor pathology, and tumor incidence. There was no difference between the control cage, the sham exposed, and the three other groups exposed to 1.4, 2.2, and 3.5 W/kg. We observed a slight increase in the number of malignant tumors at SARs higher than 1.4 W/kg. However, we observed a decrease in the number of these malignant tumors at SARs lower than 1.4 W/kg.

Our second experiment concerns the effects of microwaves on lipoperoxidation in the brain, which receives a significant amount of the energy emitted by the mobile phone and which is rich in polyunsaturated fatty acids (30% of the structure of brain cell membranes). In this experiment, we used the near-field exposure system in which only the head is exposed. We used Sprague Dawley rats that were divided into four groups: a control cage, a sham-exposed group, and two groups exposed to 1 and 4 W/kg. The animals were exposed 1 hour per day for 5 days. Twenty-four hours after the end of exposure, the brains are excised, the MDA/TBA complex is extracted, and this complex is analyzed by HPLC. Our preliminary results show no difference among the four groups. The replication of this experiment by our laboratory is in progress.

A third experiment concerns the effect of microwaves on the neurodegeneration model of rats. Acute experimental autoimmune encephalomyelitis (EAE) is used as a model of the autoimmune aspects of multiple sclerosis. This disease is characterized by inflammation of the central nervous system and infiltration by the T-cells and the macrophages. In this experiment we used the Lewis rat aged 5-8 weeks. The rats were immunized with a mixture of myelin basic protein, the mucobacterium tuberculosis, and the front adjuvant. The animals were kept 3 days after the immunization and divided into four groups: control cage group, sham-exposed group, and two groups exposed to 1 and 4 W/kg. Some other parameters that we determined daily were clinical parameters and some biomarkers concerning antibodies. The first results show that there is no action on the human response concerning antibodies. There is also no action on the EAE crisis in the beginning during the clinical peaks and at the end of the peaks—only some variations in the magnitude of the clinical score are observed.

A fourth experiment concerns the effect of microwaves on c-*fos* gene expression in rat brain cells. We used Sprague Dawley rats which we acclimatized to the rocket for 1 week. Animals were exposed once for 2 hours to SARs of 1 and 4 W/kg. We had four groups: control, exposed sham, and two exposed groups. We hope to have these results by the end of this year.

The last study we will perform concerns DNA damage, using the same techniques but different exposure systems. We will use the loop antenna and expose only the head of the rats with GSM 900 MHz at SARs of 1, 2, and 4 W/kg. This study is in progress. All of these studies are supported by France Telecom.

16

EVALUATION OF POTENTIAL GENOTOXICITY OF RADIOFREQUENCY RADIATION TECHNOLOGIES USING THE SINGLE CELL GEL ELECTROPHORESIS ASSAY: INTEGRATED LABORATORY SYSTEMS STUDY

Transcript of a paper presented by R.R. Tice
Edited by WTR Staff

This paper presents work performed by Integrated Laboratory Systems (ILS) and supported by Wireless Research, LLC (WTR). In these experiments, the single cell gel (SCG) electrophoresis assay was used to detect DNA damage. Ms. Marie Vasquez and Dr. Graham Hook also participated in the conduct of these studies.

Recent reports (Lai & Singh 1995, 1996, 1997) from University of Washington have presented evidence of the induction of DNA damage by radiofrequency (RF) fields as evidenced by DNA migration in the SCG assay. These studies were conducted using 2450 MHz RF signals from cylindrical waveguides which produce a substantial whole-body specific absorption rate (SAR); the head to body SAR ratio of is approximately 1:1. Studies conducted at Washington University using waveguides similar to those used by Lai and Singh found no evidence of the induction of DNA migration using the SCG assay (Malyapa 1998). These studies, however, used a different version of the SCG assay than that used by Lai & Singh.

The purpose of the ILS study was to evaluate the significance of these results dealing with RF fields emitted by cellular telephones. To maintain relevance to cell phone exposures, rats were exposed to voice-modulated 837 MHz analog RF signals using a loop antenna which produced a head-focused exposure, yielding a head to body SAR ratio of approximately 10:1 (see Chapter 3 in this volume; Chou et al. 1999). To reduce differences in SCG assay methodologies, the protocols developed by Lai & Singh were used in this study. One of the main thrusts of this particular study was to ensure protocol replication within and across labs; for example, Lai & Singh's protocol was adhered to from lysing conditions and the proteinase K system to electrophoresis and unwinding.

Two independent experiments were conducted with male Sprague Dawley rats 7-10 weeks old and weighing 250-300 g. There were 8 animals in each dose group. RF exposure time was 3 hours, and harvest time was 3 hours after the completion of the RF exposure. The positive control was ethylmethane sulfonate (25 mg/kg BW). Brain cells of treated and control rats were prepared according to the method developed by Lai & Singh. The electrophoretic conditions in the cell media were pH

> 13, 0.4 V/cm, 225 mA. For each animal, the extent of DNA migration was determined by scoring 100 cells (50 cells from each of two slides) using the Komet 3.1 imaging system. In the analysis, the one-tail test was used to look for an increase in DNA migration. Under the experimental conditions used, no significant increase in DNA migration in 837 MHz RF-exposed rat brain cells was observed.

References

Chou CK, Chan KW, McDougall JA, & Guy AW. Development of a rat-head exposure system for simulating human exposure to RF fields for handheld wireless telephones. Bioelectromagnetics 1999; 20(suppl 4): 75-92.

Lai H & Singh NP. Acute low-intensity microwave exposure increases DNA single-strand breaks in rat brain cells. Bioelectromagnetics 1995; 16(3): 207-210.

Lai H & Singh NP. Melatonin and spin-trap compound block radiofrequency electromagnetic radiation induced DNA strand breaks in rat brain cells. Bioelectromagnetics 1997; 18(6): 446-454.

Lai H & Singh NP. Single- and double-strand DNA breaks in rat brain cells after acute exposure to radiofrequency electromagnetic radiation. Int J Radiat Biol 1996; 4: 513-521.

Malyapa RS, Ahern EW, Bi C, Straube WL, LaRegina M, et al. DNA damage in rat brain cells after in vivo exposure to 2450 MHz electromagnetic radiation and various methods of euthanasia. Radiat Res 1998; 149(6): 637-645.

McRee DI, Chou CK, McDougall JA, Chan KW, & Guy AW. A rat-head exposure system for exposure to RF fields from handheld wireless telephones. In: GL Carlo, editor. Wireless phones & health II: State of the Science. Norwell, MA: Kluwer Academic Publishers; 2000.

17

EVALUATION OF POTENTIAL GENOTOXICITY OF RADIOFREQUENCY RADIATION TECHNOLOGIES USING THE SINGLE CELL GEL ELECTROPHORESIS ASSAY: UNIVERSITY OF WASHINGTON STUDY; COMPARISON OF ILS AND UW SCG STUDIES

Transcript of a paper presented by G.J. Hook
Edited by WTR Staff

This paper presents work conducted by Drs. Henry Lai and N.P. Singh at the University of Washington (UW) and Dr. Raymond Tice, Dr. Graham Hook, and Ms. Marie Vasquez at Integrated Laboratory Systems (ILS) as part of the program sponsored by Wireless Technology Research, LLC (WTR). The SCG assay protocol was developed by Lai and Singh (1995, 1996, 1997). The exposure system was developed by Chou et al. (1999).

In both single cell gel (SCG) electrophoresis experiments at UW, positive controls showed a significant increase in DNA migration in comparison to the cage controls and the sham controls. In experiment 1, there was a pair-wise difference at 1 W/kg. No other effects were observed, and there was no effect on the trend. This study was therefore considered equivocal. In the repeat study at ILS, no effect was seen at any dose rate, nor was there a trend (see Chapter 16 in this volume).

One of the other protocols we included that further allowed comparisons between the ILS & UW laboratories was the requirement that each laboratory produce sufficient slides to send an entire set or some portion of the set to the other laboratory to score. Looking at both percent DNA and tail moment, the slides of the two labs are reasonably close, except for that positive control. If you remember back, this is actually from the UW experiment where a very high level showed up in the positive control but that did not show up in tail moment or percent DNA. If corrections based on what a micron size would be for UW are taken into consideration for their increased magnification, then the data fall about equal.

Distribution curves of the values for each animal in the sham and in the 10 W/kg show no difference in tail length. There is no difference in the distribution here in tail length. The distribution curve of the positive effect for the UW 1 W/kg exposure shows that the effect is less of an effect due to an increase in tail length and more from a contraction at a slightly higher mean value.

There were no reproducible effects on DNA migration as measured in tail length in rat brains. We did see an equivocal response in the first test, but it was not reproduced in the second test. There was certainly no effect on the trend overall dose-response. There were no obvious effects on the distribution of damage among cells, so there were not any outlyer populations or small highly damaged sets that were hidden by the rest of the population of cells. The positive control in both experiments showed a significant effect against the sham and the cage controls. There was virtually no difference between the measured amounts of DNA migration in those two labs.

References

Chou CK, Chan KW, McDougall JA, & Guy AW. Development of a rat-head exposure system for simulating human exposure to RF fields for handheld wireless telephones. Bioelectromagnetics 1999; 20(suppl 4): 75-92.

Lai H & Singh NP. Acute low-intensity microwave exposure increases DNA single-strand breaks in rat brain cells. Bioelectromagnetics 1995; 16(3): 207-210.

Lai H & Singh NP. Melatonin and spin-trap compound block radiofrequency electromagnetic radiation induced DNA strand breaks in rat brain cells. Bioelectromagnetics 1997; 18(6): 446-454.

Lai H & Singh NP. Single- and double-strand DNA breaks in rat brain cells after acute exposure to radiofrequency electromagnetic radiation. Int J Radiat Biol 1996; 4: 513-521.

EDITOR'S NOTE: WTR TIER II

WTR's Tier II epidemiology program includes both cohort and case-control studies conducted in the context of post-market surveillance.

Prior work published by WTR has identified appropriate metrics of measuring exposure to RF radiation through the use of phone company billing records. Those metrics have been employed in the epidemiology research supported by WTR (see Chapter 18 in this volume).

On the whole, human epidemiology studies are equivocal with respect to overall risk of brain cancer following wireless phone use. However, a significant increase in neuro-epithelial brain tumors and an increase in the risk of acoustic neuromas raise legitimate questions about public health risk.

V

EPIDEMIOLOGY

18

STATE OF THE SCIENCE: RF EPIDEMIOLOGY

G.L. Carlo & R. Steffens Jenrow

Abstract

At the time that the Wireless Technology Research, LLC (WTR) program was begun in 1993, there were no epidemiological data directly relevant to the possible health effects related to the use of wireless communication instruments. As a result of both the Epidemiology Literature Surveillance and the Tier II Risk Evaluation components of the WTR program, presently there are existing data to which the public health paradigm can be applied.

This paper describes epidemiologic results from completed studies submitted for publication and coordinated through WTR, as well as other published studies identified through ongoing surveillance.

Under the WTR program, epidemiology studies are considered in the context of both post-marketing surveillance and as tools to test specific cause-effect hypotheses to supplement experimental studies. Neither of these types of studies alone can provide a solid basis for assessing the potential public health impact of wireless communication technology. Therefore, results of studies from each of these distinct disciplines must be considered and interpreted in the context of the other, and answers regarding public health risk must follow from a weight-of-evidence evaluation. All follow-up studies are addressed as recommendations made in the Tier III portion of the WTR program.

Overview of Public Health Paradigm

Since its inception, the entire WTR program has been based on a public health paradigm approach to assessing and managing public health risk. This approach, clearly outlined in WTR's Research Agenda (SAG 1994), is in contrast to more traditional regulatory or legal models for evaluating health risks. This approach essentially dictates that all studies are directed toward determining if the public's health is adversely affected and, if so, how to mitigate that risk. All activities of the program are directed toward answering one of the following four questions:

1. Is there a public health problem posed by wireless communication technology?

2. If so, what are the characteristics of that public health problem?
3. What are the appropriate corrective interventions to mitigate any identified public health risk from wireless communication technology?
4. What is the appropriate implementation strategy for those interventions?

Program Criteria

The program was framed in accordance with five criteria considered essential to assure independence and scientific rigor and to meet the scientific goals of the public health paradigm. These criteria are listed below:

1. The program must be independent of industry influence, so that the results will be acceptable to public health decision makers in government, the scientific community, and industry.
2. The program must adhere to the highest of scientific standards to guarantee scientific rigor.
3. The program must encompass a rapid trigger for public health intervention, if any adverse impact of wireless technology is discovered.
4. The program must encompass ongoing coordination with government decision-making bodies, the scientific community, and industry to ensure that the research effort is responsible to the evolving needs of these groups.
5. The program must involve significant funding to gain answers to critical public health questions in a defined time frame.

Content Areas

Based on the public health paradigm and the program criteria, four content areas were defined: Ongoing Surveillance, Risk Evaluation Research, Risk Management Research, and Scientific Outreach. The Ongoing Surveillance component of the program encompasses the daily review of new scientific information according to the public health paradigm to evaluate either a need for intervention or a need for additional information. Risk Evaluation Research involves the design and conduct of fundamental research into the possible public health impact of wireless communication technology, including dosimetry and measurement studies, in vivo and in vitro toxicology studies, epidemiology studies, and clinical studies, where applicable. The overall plan for risk evaluation research is laid out in the research agenda and encompasses a three-tiered approach:

- Tier I studies address development of appropriate exposure systems for use in in vitro and in vivo toxicology studies and the use of those systems for standardized toxicological testing. Hypotheses were tested in accordance with standard approaches to product safety evaluation.
- Tier II studies encompass epidemiological evaluations of cellular telephone users, employing appropriate measures of real-life exposures.

- Tier III studies will be conducted as needed to address mechanistic and methodologic issues arising from studies conducted under Tiers I and II which are suggestive of a public health risk.

The answers found in relation to questions 1 and 2 of the public health paradigm raise the WTR program to the level of intervention, while questions 3 and 4 of the paradigm form a pathway to intervention that improves public health.

Applying the Public Health Paradigm to Existing Epidemiological Data

As discussed earlier in this paper the Ongoing Surveillance component of the program involved the daily review of new scientific information in both epidemiology and toxicology, to evaluate either a need for intervention or a need for additional information by way of Tier III.

Risk Evaluation Research also discussed earlier involves the design and conduct of fundamental research, including epidemiology and in vivo and in vitro toxicology studies. WTR scientists have completed studies in both of these areas.

Background

Epidemiology

At the time that WTR was formed, there were no epidemiological data directly relevant to this issue. However, as a result of specific components of the WTR program, there are presently existing data to which the public health paradigm can be applied.

This paper intends to serve as an interpretational guide for application of the public health paradigm to these existing data. These data include completed studies submitted for publication and coordinated through WTR, as well as other published studies identified through ongoing surveillance. An additional section of the paper will describe other epidemiological studies of wireless technology currently in progress or unpublished to which the public health paradigm can also be applied once data become available.

Under the WTR program, epidemiology studies are considered in the context of both post-marketing surveillance and as tools to test specific cause-effect hypotheses to supplement experimental studies. Neither experimental studies nor epidemiological studies alone can provide a solid basis for assessing the potential public health impact of wireless communication technology. Therefore, results of studies from each of these distinct disciplines must be considered and interpreted in the context of the other, and answers regarding public health risk must follow from a weight-of-evidence evaluation.

To be able to apply the public health paradigm to existing data, we must first provide a brief overview of the various epidemiologic study designs. This overview will include why a particular design is chosen, how it is conducted, and its advantages and disadvantages when used for evaluating possible health effects from wireless technology.

Cohort Studies.

Cohort, or follow-up, studies compare subjects for differences in disease occurrence on the basis of their exposure, or lack thereof, to the agent under study. The exposed and unexposed cohorts are followed through time and incidence of disease or mortality is compared. Cohort studies can be retrospective, with the point of initial follow-up beginning in the past and ending at some later time, or prospective, with initiation of surveillance in the present and follow-up through disease monitoring into the future.

Cohort studies are of most value in studying diseases that occur commonly in the population but are expensive and time-consuming. Nevertheless, well-conducted cohort studies are among the most preferred of epidemiological data. Because the numbers of people using wireless technology is growing exponentially, cohort studies of this population will yield invaluable scientific information.

Case-Control Studies.

Case-control studies are useful epidemiological tools for the study of rare conditions such as brain cancer and acoustic neuromas. Because the case-control study compares past exposure in diseased and non-diseased individuals, fewer numbers of subjects are usually necessary than are required in a cohort study. However, the background prevalence of exposure (i.e., use of wireless technology) must be high enough for significant numbers of cases and controls to have been exposed.

Cross-Sectional or Ecological Studies.

A cross-sectional study looks at prevalence of disease in a cohort at one instant in time. In effect, such a study takes a snapshot of a cohort. Only current exposure information is gathered, and it is not possible in a cross-sectional study to determine whether or not the exposure preceded the effect under study. Cross-sectional studies are of limited value in assessing cause and effect but may have value in determining implementation strategies for corrective interventions aimed at identified disease risks.

An ecological study considers a population defined by one variable such as geographic location or prevalence of disease in that same location. These types of correlational studies are of limited value in determining cause and effect because there is no direct relationship between exposure and disease in individuals. These types of studies are employed most often in generating hypotheses to be tested by other epidemiological methods. Cross-sectional and ecological studies are given little weight in the assessment of public health risk.

Laying the Groundwork: WTR Working Groups

Before fundamental research under the WTR program began, several experts in epidemiology, toxicology, and radiofrequency radiation dosimetry addressed important issues relevant to wireless communication instrument users. What follows is a summary of the WTR working groups that were formed prior to the actual conduct of any WTR studies.

Exposure Assessment Working Group

Members:
Dr. George L. Carlo, Wireless Technology Research, LLC, Washington, DC
Rebecca A. Steffens, Wireless Technology Research, LLC, Washington, DC
Dr. Kenneth Rothman, Epidemiology Resources, Inc., Newton Lower Falls, MA
Donna P. Funch, Epidemiology Resources, Inc., Newton Lower Falls, MA
Dr. Chung-Kwang Chou, City of Hope National Medical Center, Duarte, CA
Dr. Arthur W. Guy, Center for Bioengineering, University of Washington, Seattle, WA
Dr. Susan Preston-Martin, Department of Epidemiology, University of Southern California, Los Angeles, CA
Dr. Jack Mandel, Department of Environmental & Occupational Health, University of Minnesota, Minneapolis, MN
Dr. Robert Morgan, Environmental Health Strategies, Inc., Redwood City, CA
Dr. Quirino Balzano, Motorola Corporate Electromagnetic Research Laboratory, Ft. Lauderdale, FL

This working group studied a number of cellular technology and human exposure issues in order to provide guidance in epidemiological studies. These issues follow:

- Descriptions of cellular technology, its place in the electromagnetic spectrum, and the specific absorption rate (SAR) as a dosimetric parameter used to measure exposure;
- Population exposure which involves length and duration of telephone calls, as well as other types of radiofrequency (RF) radiation exposures;
- Occupational exposure;
- Use of billing records;
- Meaning of cumulative use per month;
- Cell site density and power;
- Alternative exposure metrics; and
- Tissue-specific exposure (Rothman et al. 1996a).

Tissue Type Working Group

Members:
Dr. George L. Carlo, Wireless Technology Research, LLC, Washington, DC
Rebecca A. Steffens, Wireless Technology Research, LLC, Washington, DC
Dr. Kenneth Rothman, Epidemiology Resources, Inc., Newton Lower Falls, MA
Dr. Arthur W. Guy, Center for Bioengineering, University of Washington, Seattle, WA
Dr. Susan Preston-Martin, Department of Epidemiology, University of Southern California, Los Angeles, CA
Dr. David Savitz, Department of Epidemiology, University of North Carolina, Chapel Hill, NC

Careful consideration was given to the different tissues of the head, neck, and upper body that may be affected by exposure to low-power devices such as handheld cellular telephones. Using a SAR diagram showing the configuration of a handheld portable "flip" cellular telephone to the head during a normal phone call and the SAR recorded at several points during this exposure, the Tissue Type Working Group was able to identify five distinct tissue types. The highest SARs (1.1 and 0.9 mW, respectively) are seen in front of the ear and directly over the ear, while the lower SARs are along the cheek and directly behind the ear. These observations, along with the unique characteristics of several types of cancers and/or tumors and other factors such as biological plausibility, latency considerations, prevalence, severity, prognosis, and so on, led the working group to determine that these tissues warranted further examination in terms of RF exposure and should be considered for subsequent case-control studies. These are gliomas, meningiomas, acoustic neuromas, salivary gland tumors, and adult onset leukemia.

Data in Existence

The existing epidemiological data are presented in table form with a column identifying whether any intervention was required after evaluating the results using the public health paradigm approach. Table 1 identifies epidemiology studies completed within the WTR program, Table 2 includes epidemiology studies identified through ongoing literature surveillance, and Table 3 lists ongoing or planned epidemiological research.

Investigator/ authors	Study type/size	Outcome	Wireless technology	Results	Intervention?	More research?
Rothman et al., Epidemiology Resources, Inc., 1996 (U.S.)	Questionnaire-based pilot study on 5,000 phone users	Assess utility of information from billing records; assess questionnaire validity; evaluate handedness as a surrogate for laterality of phone use.	Analog phones	Billing records are useful; high correlation between answers to questionnaire and billing records; handedness was not associated with laterality of phone use.	N	Y
Rothman et al., Epidemiology Resources, Inc., 1996 (U.S.)	Record-based cohort of 250,000 cellular telephone users	Overall mortality among users of handheld versus non-handheld telephones.	Handheld portable versus non-handheld	Age-specific rates were similar; large-term increase in mortality is not associated with RF exposure over the short term.	N	Y
Dreyer et al., Epidemiology Resources, Inc., 1998 (U.S.)	Record-based cohort of 496,460 cellular telephone users	Cause-specific risk of death among users of handheld versus non-handheld telephones.	Handheld portable versus non-handheld	Most striking increase among handheld cellular telephone users was due to motor vehicle accidents. Deaths from brain cancer were greater than expected, but there were only 2 cases, too few to draw any firm conclusion.	N	Y

Table 1 (continued on next page). Epidemiology studies completed within the WTR program.

Investigator/ authors	Study type/size	Outcome	Wireless technology	Results	Intervention?	More research?
Muscat et al., American Health Foundation, 1999 (U.S.)	Case-control; 90 newly diagnosed cases of AN, and 86 age-matched controls with acute conditions. All cases were histologically confirmed.	Acoustic neuroma	Self-reported phone use 1)Frequency of use, 2) Duration of use (years) and 3) Cumulative frequency (frequency X duration).	No association between AN risk and frequency or cumulative frequency. An OR of 1.5 was associated with 6 or more years of cell phone use $p>0.05$. A nonparametric multivariate regression model showed that the risk of AN decreases with short term usage ($p<0.05$), but increases with 6 or more years of use $p>0.05$.	N	Y
Muscat et al., American Health Foundation, 1999 (U.S.)	Case-control; 465 patients with newly diagnosed brain cancer and 421 age-matched controls	Brain tumors (meningiomas excluded)	Self-reported phone use 1)Frequency of use, 2) Duration of use (years) and 3) Cumulative frequency (frequency X duration)	No association was observed between brain cancer risk and cell phone use. A histologic-specific analysis, showed a significant association for 34 patients with ganglioneuromas/ ganglioneuroblastomas (14/34) who were cell phone users. No association with frequency or duration variables.	N	Y

Table 1 (continued from previous page). Epidemiology studies completed under the WTR program.

Investigator/ authors	Study type/size	Outcome	Wireless technology	Results	Intervention?	More research?
Mann et al., 1997 (Germany)	Cross-over design/ 22 healthy subjects	Nocturnal hormone profiles of growth hormone, cortisol, luteinizing hormone, and melatonin	GSM mobile phone system, 900 MHz, pulsed with 217 Hz, average power density 0.02 mW/cm^2 at 40 cm distance	No significant effects except for a slight, transient elevation in cortisol production	N	N
Mild et al., 1998 (Sweden, Norway)	Cross-sectional/ 9,114 Swedes and 3,200 Norwegians	Subjective symptoms (headaches, fatigue, warmth behind/on the ear, burning of the skin). Covariates considered were gender, age, VDT work, and an index of psychosocial work load.	GSM and NMT900 mobile phones	The prevalence of fatigue and headaches was similar for GSM and NMT users in Norway, whereas it was lower for GSM users in Sweden. In both countries, GSM users reported a lower frequency of feeling of warmth on/behind the ear compared to NMT users. There was a positive relationship between prevalence of symptoms and number of calls per day and total number of minutes using the phone per day for both transmitter systems.	N	N
Wagner et al., 1998 (Germany)	Cross-over design experiment/ 22 males	Sleep	GSM mobile phone system, 900 MHz, pulsed with 217 Hz, average power density 0.2 W/m^2 at 40 cm distance	Sleep-onset latency and duration of REM sleep tended to be reduced while the occurrence of first REM episode tended to be prolonged with exposure to the RF field. None of these changes were statistically significant.	N	N

Table 2 (continued on next page). Epidemiology studies identified through WTR literature surveillance.

Investigator/ authors	Study type/size	Outcome	Wireless technology	Results	Intervention?	More research?
De Seze et al., 1998 (France)	Cross-over design experiment/ 18 males	Anterior pituitary hormones (ACTH, TSH, GH, prolactin, luteinizing hormone, and follicle stimulating hormone). Covariates considered were age, seasonal variation, smoking status, and time of exposure.	GSM cellular phones (890-900 MHz, maximum power 2 W)	No persistent or cumulative changes in the secretion of the anterior pituitary hormones were detected. Only a 21% decrease in TSH concentration was detected during the radiotelephone exposure period, which remained within the physiologic range and recovered fully during the post-exposure period.	N	N
Braune et al., 1998 (Germany)	In vivo experiment/ 7 males, 3 females	Blood pressure, heart rate, capillary perfusion, and subjective well-being.	GSM, 900 MHz, 2 W, 217 Hz frame repetition rate	Exposure to the phone caused an increase in sympathetic efferent activity with increases in blood pressure between 5 and 10 mm Hg, probably due to more pronounced vasoconstriction. No statistically significant effects on subjective parameters were found.	N	N
Freude et al., 1998 (Germany)	In vivo experiment/ 16 males	Human slow brain potentials	GSM, 916.2 MHz, 350 mW average power	EMF exposure produced a significant decrease of slow brain potentials at central and temporo-parieto-occipital brain regions, but not at the frontal region.	N	N

Table 2 (continued from previous page). Epidemiology studies identified through WTR literature surveillance.

Investigator/authors	Study type/size	Outcome	Wireless technology	Date results expected
National Cancer Institute (U.S.) Inskip, Linet, et al.	Multi-site hospital-based case-control; 700 case/controls	Glioma, meningioma, and acoustic neuromas	Primarily analog phones	Data collection complete and analysis is underway
International Agency for Research on Cancer (IARC) (Multinational) Cardis	Two stage, multinational case-control study Pilot study conducted in Australia, Canada, France, Germany, Israel, Italy, New Zealand, Sweden, and the U.K.	Pilot study investigated prevalence of phone use over the last 10 years, possible sources of information for estimating history of phone use, and calculated statisical power of the main study to follow. Main study will examine tumors of the acoustic nerves and salivary glands, menigiomas, and gliomas.	European technology	Pilot study completed in 1998; Main study is expected to begin in late 1999 or early 2000 and to end by 2004
European Commission (Multinational) McKinlay et al.	Several different types	Brain tumors, acoustic neuromas, salivary gland tumors, adult-onset leukemia, headaches, inner ear, neurobehavioral issues, and pineal melatonin secretion.	Digital	Planned

Table 3 (continued on next page). Additional worldwide ongoing or planned epidemiologic research.

Investigator/authors	Study type/size	Outcome	Wireless technology	Date results expected
Hardell & Mild (Sweden)	Case-control	600 cases and 600 controls of histologically confirmed brain tumor patients	European technology	Ongoing, but preliminary results show some evidence of laterality of phone use with tumors
Johansen (Denmark)	Nationwide cohort, 800,000 subscribers	Central nervous system tumors	European technology	Ongoing
Juutilainen (Finland)	Computer modelling, in vivo, in vitro, epidemiologic feasibility study	Health risks possibly associated with RF exposure	European technology	Ongoing
Stough (Australia)	Not available	Effects of mobile phone use on memory, concentration, and problem-solving abilities	Not available	Planned
World Health Organization (Multinational)	The International EMF Project	Health and environmental effects of exposure to static and time-varying static, ELF, and RF fields	European technology	Ongoing

Table 3 (continued from previous page). Additional worldwide ongoing or planned epidemiologic research.

Discussion: Interpretation of Results Obtained in Tiers I & II According to Public Health Paradigm

Tier III studies as defined earlier should be conducted to address mechanistic and methodologic issues arising from studies conducted under Tiers I and II of the WTR program which were suggestive of a public health risk from wireless technology. Since both epidemiological and toxicological data now exist with which public health decisions can be made, this follow-up phase should be defined and implemented. This phase will still encompass many parts of the original WTR program; however, it will be streamlined to reflect the information obtained in Tiers I and II and the decisions made as a result of this information.

Ongoing surveillance and research will still make up the primary components of the next phase of RF research. Ongoing surveillance should include both passive surveillance and active surveillance. Passive surveillance is the ongoing monitoring of existing literature in the following areas: epidemiology, toxicology, and exposure systems. It also consists of following any relevant government actions as well as anecdotal reports of health effects from the field. Active surveillance will include follow-up of the record-based epidemiological cohort of analog phone users developed under Tier II. Based on information also obtained from Tier II a record-based cohort of digital users for prospective follow-up should be created. This could be done by

1. Implementing the Lifestyle study protocol already accepted under the first phase of the WTR program;
2. Creating a new cohort by using information obtained from warranty cards filled out whenever a digital phone is purchased; or
3. Creating a new cohort from billing records of digital phone users similar to the methods already developed for analog users.

If data from either of these cohorts should indicate a health risk, a decision tree mechanism is already in place that will provide a rapid trigger for intervention. This intervention would include tissue-specific case-control studies and/or exposure-specific cohorts. Finally, under Ongoing Surveillance will be a system for monitoring newly developed technology and testing it against the Tier I and Tier II results already obtained.

Under the Research component of Tier III will be follow-up studies conducted as a result of information obtained in Tier I and Tier II. The following studies are recommended:

<u>Epidemiology</u>

1. Neuroblastoma follow-up;
2. Adult-onset leukemia case-control study;
3. Salivary gland tumor case-control study; and
4. Meningioma case-control study.

Toxicology
1. Micronucleus assay follow-up, and
2. Transgenic mouse follow-up.

Any RF program should also continue to follow and participate in ongoing research being done by other groups including, among others:

1. US National Cancer Institute;
2. International Agency for Research on Cancer (IARC);
3. World Health Organization International EMF Project;
4. European Commission RF Program;
5. COST 244; and
6. Bioelectromagnetics Society (BEMS).

Conclusion

There is a new wireless customer every 2.25 seconds. From a public health point of view, it is imperative to simultaneously accomplish surveillance, tracking, monitoring, and the ability to respond quickly so that consumers can be protected. At the same time, it is necessary to conduct research that is rigorous, is peer reviewed, and will stand the test of time. Epidemiological studies and definitive animal studies take years to complete, which in some sense limits their value to rapidly changing technologies. Epidemiology studies also have disease latency limitations even with static technologies. Meanwhile, wireless technology is not only rapidly growing, but it is rapidly evolving. Therefore, in addition to traditional research, there is a need to address consumer protection. One way to accomplish this is to apply the public health paradigm to information that is generated in a variety of ways in the interest of protecting the public.

References

Braune S, Wrocklage C, Raczek J, Gailus T, & Lucking CH. Resting blood pressure increase during exposure to a radio-frequency electromagnetic field (Research Letter). The Lancet 1998a; 351: 1857-1858.

Braune S, Wrocklage C, Raczek J, Gailus T, & Lucking CH. Radio-frequency electromagnetic field from mobile phones (Author's reply). The Lancet 1998b; 352: 577-578.

de Seze R, Fabbro-Peray P, & Miro L. GSM radiocellular telephones do not disturb the secretion of antepituitary hormones in humans. Bioelectromagnetics 1998; 19: 271-278.

Dreyer NA, Loughlin JE, & Rothman KJ. Cause-specific mortality in cellular telephone users. 1998 (submitted for publication).

Freude G, Ullsperger P, Eggert S, & Ruppe I. Effects of microwaves emitted by cellular phones on human slow brain potentials. Bioelectromagnetics 1998; 19:384-387.

Funch DP, Rothman KJ, Loughlin JE, & Dreyer NA. Utility of telephone company records for epidemiologic studies of cellular telephones. Epidemiology 1996; 7: 299-302.

Juutilainen J & de Seze R. Biological effects of amplitude-modulated radiofrequency radiation. Scand J Work Environ Health 1998; 24: 245-254.

Mann K, Wagner P, Brunn G, Hasssan F, Hiemke C, & Roschke J. Effects of pulsed high-frequency electromagnetic fields on the neuroendocrine system. Clinical Neuroendocrinology 1998; 67: 139-144.

Mild KH, Oftedal G, Sandstrom M, Wilen J, Tynes T, Haugsdal B, & Hauger E. Comparison of analogue and digital mobile phone users and symptoms: A Swedish-Norwegian epidemiological study. 1998. (This report can be downloaded from the Swedish National Institute for Working Life website: http://www.niwl.se.)

Muscat J. Cellular telephone use and risk of acoustic neuroma. 1999 (submitted for publication).

Muscat J. Cellular telephone use and risk of brain cancer. 1999 (submitted for publication).

Muscat J. Wireless phone use and the risk of primary brain cancer. In: GL Carlo, editor. Wireless phones & health II: State of the science. Norwell, MA: Kluwer Academic Publishers; 2000.

Repacholi MH. Low-level exposure to radiofrequency electromagnetic fields: Health effects and research needs. Bioelectromagnetics 1998; 19: 1-19.

Rothman KJ, Chou CK, Morgan R, Balzano Q, Guy AW, Funch DP, Preston-Martin S, Mandel J, Steffens R, & Carlo G. Assessment of cellular telephone and other radio-frequency exposure for epidemiologic research. Epidemiology 1996a; 7: 291-298.

Rothman KJ, Loughlin JE, Funch DP, & Dreyer NA. Overall mortality of cellular telephone customers. Epidemiology 1996b; 7: 303-305.

Scientific Advisory Group (SAG) on Cellular Telephone Research. Potential public health risks from wireless technology: Research agenda for the development of data for science-based decisionmaking. Washington, DC: SAG; 1994.

Wagner P, Roschke J, Mann K, & Frank C. Human sleep under the influence of pulsed radiofrequency electro magnetic fields: A polysomnographic study using standardized conditions. Bioelectromagnetics 1998; 19: 199.

19

EPIDEMIOLOGICAL STUDY ON THE USE OF CELLULAR TELEPHONES AND THE RISK FOR BRAIN TUMORS

K. Hansson Mild & L. Hardell

The principal investigator of this study is Dr. Lennart Hardell. He was unable to attend the Colloquium, therefore Dr. Kjell Hansson Mild gave the presentation. This paper is based on a transcript from his talk. Other researchers involved in this study are Åsa Näsman, Department of Oncology; Dr. Anneli Påhlson, Department of Neurology, Örebro Medical Center, Örebro; and Dr. Arne Hallquist, Department of Oncology Karolinska Institute and Stockholms sjukhem, Stockholm, Sweden. This study has recently been published in the International Journal of Oncology (1999; 15: 113-116).

The use of cellular telephones has increased dramatically during the 1990s across the world. In the 1980s the analogue NMT system was used in Europe, whereas the digital GSM system introduced in early 1990s is now the preferred system. Case reports of brain tumors in users initiated this case-control study on brain tumors and use of cellular telephones. Other exposures were also assessed. We collected cases in the Uppsala-Örebro region between the years 1994-1996 and in the Stockholm region between the years of 1995-1996. The subjects ranged in age from 20 to 80 years old. The criteria for inclusion in the study included the subjects having been diagnosed with a histopathologically verified malignant brain tumor. In 1996 subjects with benign tumors were also included in the Stockholm region. The cases should be living at the time of interview, thus no proxy interviews were used. The total number of cases originally identified and fulfilling these inclusion criteria was 270, but for medical reasons we were only able to use 233 cases.

As a control population, we used two persons for each case matched for gender, age, and region where they were living. They were selected from the Swedish Population Register. Questionnaires were sent out to the cases and controls and were followed up with telephone interviews. A total of 217 cases sent back the questionnaire and completed the telephone interview. However, when scrutinizing the medical records after permission of the case, it turned out that 8 of these cases had a recurrent brain tumor. They were excluded from the analysis along with their 14 matched controls. The analyses were thus based on answers from 209 cases and 425 controls. The distribution rate between male and female participants was about the same. The mean age for cases and controls was 50 years. The histopathological distribution of the tumors of the cases in the study is shown in Table 1. We were able to obtain medical records for 197 of the cases.

Histopathology	Number of cases
Malignant	136
Glioblastoma	48
Astrocytoma	46
Oligodendroglioma	19
Ependymoma	3
Mixed glioma	16
Malignant NUD	4
Benign	62
Meningioma	46
Acoustic neuroma	13
Other benign	3

Table 1. Histopathological distribution of brain tumors for 197 patients. One case had both ependymoma and acoustic neuroma.

The same percentage of cases and controls had used a mobile phone: 37.3% and 37.9%, respectively. All subjects were asked about the number of minutes per day and how many years they had used their phone, what type of phone they used, and what hand they used to hold their phone during calls. In addition, we specifically asked them about use of hands-free devices because these were excluded from the exposure pattern. All exposure within 1 year prior to diagnosis was excluded for both cases and the matched controls. We estimated the total number of hours they used their phones, which we deemed our "dose measure" in the study. The mean number of use hours was 511 for the cases (range of 8-9000) and 428 for the controls (range of 8-6304).

A total number of 78 cases and 161 controls reported use of a mobile phone. Thus, no increased risk was found (see Table 2). Then we divided the use according to type of mobile phone: the digital GSM system and the analogue NMT system. The NMT is the system that has been in use for quite some time. Again, we did not observe any increased risk, as shown in Table 2.

	Cases/controls	Odds ratio	Confidence interval
Latency period > 1 year			
Total	78/161	0.98	0.69-1.41
< 136 hours	38/76	1.01	0.64-1.58
≥ 136 hours	40/85	0.96	0.61-1.52
GSM	44/92	0.97	0.61-1.56
< 88 hours	20/45	0.93	0.49-1.79
≥ 88 hours	24/47	1.00	0.56-1.80
NMT	52/106	0.94	0.62-1.77
< 224 hours	28/51	1.05	0.62-1.77
≥ 224 hours	24/54	0.83	0.47-1.47

Table 2. Odds ratios and 95% confidence intervals for exposure to cellular phone. Latency period > 1 year. Dose-response calculations were based on median number of hours for exposure among controls.

We also studied 5- and 10-year latency periods (see Table 3). We saw no increased risk depending on tumor induction time or a dose-response effect.

Next we researched how the subjects had been using their phones. We observed that, of the cases, 28% used their left ear, 57% used their right ear, and about 15% used both ears during calls. These numbers were about the same for the controls: 30% left ear, 61% right ear, and 9% both. It is of interest to note that about 10% of the population are left handed, so a certain number of people will use the mobile phone with their left hand but the majority of left-handed users use their right hand.

	Cases/controls	Odds ratio	Confidence interval
Latency period > 5 years			
Total	34/69	0.83	0.49-1.42
< 424 hours	16/34	0.87	0.42-1.77
≥ 424 hours	18/35	0.80	0.40-1.61
Latency period > 10 years			
Total (NMT)	16/26	1.20	0.56-2.59
< 968 hours	9/13	1.32	0.47-3.70
≥ 968 hours	7/13	1.06	0.33-3.40

Table 3. Odds ratios and 95% confidence intervals for 5- and 10-year tumor induction (latency) periods.

	Cases	Odds ratio	Confidence interval
Temporal or occipital	25	1.47	0.78-2.77
Right side & right ear	8	2.45	0.78-7.76
NMT (ever)	7	2.56	0.72-9.12
GSM (ever)	3	1.23	0.26-5.75
Left side & left ear	5	2.40	0.52-10.9
NMT (ever)	5	2.10	0.46-9.57
GSM (ever)	1	0.78	0.07-8.88

Table 4. Odds ratios and 95% confidence intervals for exposure to cellular phones according to tumor localization in relation to ear (side) used for phone. Latency period > 1 year.

We then focused on the people that portrayed a side-of-use preference. We analyzed the risk associated with a subject having a tumor located on the same side of their head as they use their phone (see Table 4). The total number of cases with a tumor on the right side of the head in the temporal or occipital lobe (temporoparietal localization included) and use of the mobile phone on the same side was 8 yielding an odds ratio of 2.45 (95% confidence interval = 0.78-7.76). Similarly, 5 cases with a left-sided tumor in these lobes reported use of a cellular phone on the left ear, yielding an odds ratio of 2.40 (95% confidence interval = 0.52-10.9). The increased risk was only found for NMT users, although only 1 case had used only GSM.

This can be regarded in two ways. Since the GSM phone was introduced in the early 1990s, we were not expecting to see many GSM users among the 63 cases with a tumor in the temporal (n=48), occipital (n=7), or temporoparietal (n=8) lobe. If there was an effect with the phone, the NMT phones that are being used in Sweden have a power output of 1 W. Conversely, GSM phones have a regulated output power in the milliwatt range, a rather large difference in output power in comparison to NMT phones. Therefore, there would also be a direct influence on the specific absorption power (SAR) based on the type of phone.

If we specifically look at the 13 cases of acoustic neuroma, 5 of these 13 diagnosed cases had used a mobile phone. This yields an odds ratio of 0.78 (95% confidence interval = 0.14-4.20). Only one case had a tumor on the same side as the exposure.

In conclusion, future research should focus on the anatomical area closest to the ear used for the mobile phone. We are currently continuing to collect cases and controls to further study the use of cellular phones and the risk for brain tumors, if such a risk exists. Presently we have interviewed about 1,000 cases and a similar number of controls. Our goal is to have about 1,500 cases and an equal number of controls in this study. We hope to have more data by this time next year.

20

WHO'S KEY RF RESEARCH REQUIREMENTS FOR BETTER HEALTH RISK ASSESSMENTS

Transcript of a paper presented by M. Repacholi

Edited by WTR Staff

The World Health Organization (WHO) has specific research needs that are essential to improving health risk assessments. A program, called the International EMF Project, has been developed within WHO to look at the health and environmental effects of exposure to the electromagnetic field spectrum from static to 300 GHz fields. Instead of reviewing a database and observing what is needed, the Project's approach to surveying the research was to examine the content of a database needed by WHO and the International Agency for Research on Cancer (IARC), to assess carcinogenic and other health hazards, and then to conduct essential research to fill the gaps. These gaps are mainly in the form of epidemiology and toxicology studies. In order for WHO to compile a database that will provide adequate health risk assessments, we identified research that was suggestive of a health effect, but not substantiated. Now, the research on these health effects needs to be confirmed or extended.

Databases for the epidemiological, human studies, and animal studies are the most useful. They are major indicators of health risk primarily because you do not have to extrapolate the results. Instead, you can take results of those studies and directly apply them to health risk assessments. In vitro laboratory studies, where there is a battery of tests like the ones that have been conducted on DNA in order to find out about micronucleus formation, chromosome aberration, or mutations, are extremely useful if you can tie the results to an in vitro effect. In vitro studies are also especially useful when teamed with in vivo results and used for health risk assessments. Then they can provide summations of mechanisms of interaction and allow for the extrapolation of in vivo results to humans. Similarly, when studies are performed, it is preferable when the outcome you are looking for is able to be achieved by humans, in laboratory situations. This makes the data much more reliable than what you can get from epidemiology or animal studies. Even though you cannot look for cancer in human laboratory studies, there are a number of other non-cancerous effects that can be found.

When looking at animal studies, it should be known that one animal study alone cannot be used for health risk assessments. The studies must be replicated in order to back up the results. Animal studies must be 2-year bio-assays or promotion studies using initiators. Also, many different types of animal models need to be utilized because it is possible to get species-specific effects such that what might occur in one animal model might not occur in another. In this case, the results need to be duplicated in at least two different animal models to give confidence that they can be extrapolated to humans. Verified consistence of positive results is another point that is essential to complete. The converse of this is true, where negative results may not be helpful to risk assessments due to poor methodology. Overall, if a result is negative, you must look at the methodology that was used to see if a positive result could have been achieved.

Many people have been involved in WHO's health risk assessments and have helped compile an agenda that WHO would like to have completed. This agenda will provide the current research information that is sufficient for much improved health risk assessments. Currently, WHO has a research agenda that is promoting research with funding agencies in order to complete the agenda. The key is that we need high quality research, and WHO prefers the use of a standard protocol. This way, data can be combined through meta-analysis which will then increase the power and significance of the studies. Quality insurance procedures in the protocol are also necessary and must be maintained during the study so there is confidence that the study is conducted in a reasonable manner.

WHO's current research priorities in radiofrequency (RF) radiation are mobile telephones because their use is increasing so rapidly. WHO has estimated that there will be about 700 million users worldwide in the next 3 to 4 years. WHO's interest is specialized mainly because it is much more efficient and economical for developing countries to use mobile telephone technology and bypass the landline systems. In turn, it is in WHO's interest to ensure that any health effects from mobile phones that may be found will not occur in such developing countries.

Since the technology is increasing so rapidly, we need to complete these studies now. In 5 to 10 years virtually everyone will be using a mobile telephone and we will lack control populations. It is necessary that we promptly conduct large studies with large numbers of cases and controls to find if there is any impact. Even if there is only a small consequence, it could be a major health impact because of the large numbers of people involved. As a parallel, the use of diagnostic ultrasound was a technology that came about 20 years ago. This technology has since been used enormously. Originally, public health authorities were concerned if these diagnostic units could cause any adverse health effects to the fetus. By the time studies were done, virtually every pregnant mother was using an ultrasound. Today we will never know from the epidemiology studies performed if there is any impact on the fetus because everyone is using the technology. We do not want this same thing to happen with mobile telephones.

It is also in WHO's interest to study health risks of workers exposed to high levels of RF. It is very important to note, though, that conducting epidemiology studies around mobile telephone base stations or radio and telephone transmitters is not an accurate methodology because you cannot distinguish between the exposed and control groups due to the low-level fields involved.

It is also important to note that there are emerging technologies coming out which have not been researched but on which research needs to be performed. These involve security devices and anti-theft devices, both of which provide quite significant electromagnetic fields. Research recommendations for these devices will soon be provided by WHO.

A major part of the research that needs to be compiled is dosimetry. Quality dosimetry is essential, and as we have seen in recent years, the precision of dosimetry is improving all the time. To provide updated information, we held a workshop at the UK's National Radiation Protection Branch (NRPB) and looked at exposure metrics and dosimetry for EMF epidemiology. This has just been published in *Radiation Protection Dosimetry* and provides a very good collection of papers and reports on what is needed for solid epidemiological assessments. A meeting in Zurich was conducted this past March, in which WHO participated, for the mobile manufacturers. This meeting was designed to assist the manufacturers select labs that could be submitted to the 5th framework project of the European Commission. Our responsibility was to identify labs that were qualified through experience and ability to perform research on EMF. WHO recommends that if a manufacturer is going to fund research that it must be done through an independent panel. Because a lot of the research funding is coming from the manufacturers, an independent panel ensures the researchers' independence.

Psychosocial effects also need to be studied within the International Project on EMF. There are people who experience stress reactions from exposure to fields or from their perceptions that their reactions are caused by exposure to these fields. WHO has experience with this from the Chernobyl accident where a lot of radioactive material was spread over living areas in three countries: Russia, Ukraine, and Belarus. People living in these highly contaminated areas became extremely stressed about the possibility of themselves or their children developing adverse health effects from the radiation. EMF hypersensitivity has been established, especially in Nordic countries, where some people tend to have skin rashes, headaches, or a number of other symptoms. More recently, in Switzerland there was a problem with a short wave transmitter that was assembled, where people were claiming sleep disturbances. The study suggested that while the transmitter was on, the effect occurred; when it was off, the effects did not occur. In a case such as this, it is necessary to conduct follow-up studies, especially in the laboratory, where you can clearly control for the exposure and be precise about the procedures of the study. In all, there is a need for psychosocial studies. The effects would not be controlled through standards; they can be controlled through education.

WHO has set a schedule of health risk assessments that will be conducted both by IARC and by WHO. IARC will be focusing on RF fields using the same process that the US National Institute of Environmental Health Sciences (NIEHS) has done for electromagnetic fields. It will be classifying RF as a possible or probable carcinogen. IARC is a specialized agency of WHO which has looked at over 800 physical and chemical agents using this classification system which makes them qualified to do this task. WHO will establish task groups to look at non-cancer health risks, which will be done a year after the completion of the IARC study. Then results of these studies will be published in the IARC's monographs and WHO's environmental health criteria.

Lastly, WHO has a series of fact sheets for recommending research at the time we perform our reviews, which is part of our educational program. We take the conclusions of the international scientific reviews and we have summary working group reports that are published. The results and conclusions are then used by WHO to be put into lay information and are displayed on our website. Currently, the fact sheets exist in 10 languages.

21

WIRELESS PHONE USE AND THE RISK OF PRIMARY BRAIN CANCER

J.E. Muscat

The increase in public concern over the safety of wireless phones has resulted in an international research effort to determine the acute and chronic effects of cell phone use. One particular fear is that the radiofrequency (RF) electromagnetic energy from cell phones causes brain cancer. Both laboratory and epidemiologic studies will provide crucial data on the possible carcinogenic effects of cell phone use. Risk assessment is ultimately based on the results of well-conducted studies in human populations. However, epidemiologic studies of brain cancer are difficult to conduct because the tumor is uncommon. One approach is to conduct a multi-center case-control study where newly diagnosed patients with brain cancer are queried about their cell phone usage. We took this approach to assess the risk of brain cancer and lifetime history of cell phone use.

Methods and Results

Adult patients were interviewed between 1994 and 1998 in hospitals in New York, Rhode Island, and Massachusetts using a structured questionnaire. A definition of each type of cell phone (handheld, bag, car) and of a cordless telephone was read to the patients. For each handheld cell phone used "on a regular basis," information was obtained on the dates of use, minutes/hours used per month, year of first use, manufacturer, and reported average monthly bill. The questionnaire also included items on other risk factors for brain cancer.

Cases were patients with a newly diagnosed and histologically confirmed primary brain tumor (ICD9 CM codes 191.0-191.9). Controls were patients without brain cancer admitted for acute or chronic conditions. Controls were matched to cases by age (±5 years), gender, race, and month of admission.

The risk of brain cancer was examined in relation to three measures of cell phone use: hours per month, years of use, and cumulative hours. The cumulative measure of cell phone exposure was obtained by multiplying the hours of cell phone use by the number of years of use.

The study included 469 cases and 422 controls. The mean age was 49.6 years in cases and 50.1 years in controls. The mean years of education was 14.8 in cases and 15.1 years in controls. Ninety-eight percent of all subjects were white. It is useful to examine the usage of cell phones in the controls in order to determine if the their usage is similar to that in the general population. In controls, cell phones were used more frequently by men than women and by subjects under age 60. Nine percent of controls aged 60-69 were cell phone users and no controls over age 70 were cell phone users. Cell phone use was more common in sales than in other occupations such as professional, clerical, and jobs involving physical labor.

Eighteen percent of controls and 14% of cases used handheld cell phones. Information on years of cell phone use was provided by all cases and controls. There was little difference in cell phone use by control diagnostic category, with the exception of malignant neoplasms. Six percent of controls diagnosed with cancer were cell phone users; this difference likely reflects their older age at diagnosis compared to other control patients (58.4 years vs. 47.5 years).

. Overall, cell phone handedness data were obtained for 700 of the 891 (78.6%) subjects. Self-reported billing information was requested from 102 of the 142 subjects (71.8%) that used cell phones. Information on the frequency of cell phone use was provided by all subjects except 4 cases and 1 control. All but 4 subjects that used cell phones were current users at the time of diagnosis. Four cases used cell phones for 6 months; the rest of the users reported 1 or more years of use. The crude odds ratio (OR) for brain cancer and cell phone use was 0.74 (95% CI:0.5, 1.1). The Mantel-Haenszel (MH) OR, adjusted for participating institution, was 0.79 (95% CI: 0.6, 1.1). The correlation between the frequency of the most recently used cell phone and the associated monthly bill was 0.58 ($P<0.01$).

Among cell phone users, there was no significant case-control difference in the mean number of hours used per month (11.4 in cases and 16.5 in controls, $P=0.27$). Use of a cell phone was associated with a nonsignificant decreased risk of brain cancer for both infrequent and frequent users. For instance, the OR in the most frequent users (> 10.1 hours per month) was 0.6 (95% CI: 0.3-1.2).

There was no significant case-control difference in the mean number of years of cell phone use (2.9 years in cases and 2.7 years in controls, $P=0.66$). The OR was 0.7 (95% CI:0.3, 1.3) for subjects using cell phone for 3 or more years. There was a moderate correlation between the frequency and years of cell phone use ($r=0.34$, $P<0.01$). The risk of brain cancer was below unity for all levels of cumulative use.

As an alternative method for examining the risk of brain cancer, cubic spline analysis showed that the risk of brain cancer was below unity for both low and high levels of lifetime use. The nonparametric analysis for exposure measured by frequency and duration of use was also similar to logistic regression findings.

In cell phone users, 86% of cases and 85% of controls extended the antenna during the call. Eighty-eight percent of all cell phone users used analog phones. There were no case-control differences in the percent that used digital phones.

In cell phone users, the most common tumor site was the frontal cortex

Category*	N**	Odds ratio (95% CI)	Matched odds ratio (95% CI)
Glioma	417 (353)	0.6 (0.4, 0.9)	0.7 (0.5, 1.1)
Neuroepitheliomatous	34 (34)	2.6 (1.2, 5.4)	1.7 (0.7, 4.9)

* Excludes cases with other or unspecified histology.
** Number in parentheses indicate sample size of cases in matched analysis.
Table 1. Risk of brain cancer and handheld cell phone use by two major histologic categories.

followed by the temporal cortex and parietal cortex. For the analysis of site-specific odds ratio, the entire control series was used as the comparison group. With the exception of the frontal cortex, the risk of brain cancer in all other sites was below 1.0. For the frontal cortex, the OR was 1.1 (95% CI:0.7, 1.9).

The Kappa statistic for testing a chance association between cell phone handedness and tumor laterality was 0.26 (95% CI:0.0, 0.5). In contrast, the Kappa statistic for testing the association between the hand used to hold a landline telephone and tumor laterality was 0.04 (95% CI:-0.1, 0.2). For tumors located behind the ear (the temporal cortex), the Kappa statistic was -0.24 (95% CI: -0.7, 0.2).

The risk of brain cancer was examined by two histologic groups: glioma (M9380-9481) and neuroepitheliomatous neoplasms (M9490-9523; predominantly ganglioglioma and ganglioneuroblastoma). Using the entire control series, the OR from unconditional regression was 2.6 (95% CI:1.2, 5.4) for neuroepitheliomatous tumors and 0.6 (95% CI:0.4, 0.9) for gliomas (see Table 1). For both histologic groups, there was no dose-response in the OR. In matched analysis, the OR was 1.7 (95% CI: 0.8-3.6) for neuroepitheliomatous tumors and 0.7 (95% CI:0.5, 1.1) for gliomas. There was also no dose-response in the matched ORs. None of the cases with neuroepitheliomatous tumors reported using a digital phone.

Discussion

This study shows that the risk of malignant brain cancer is unrelated to the use of handheld cell phones. The findings of no effect were confirmed in nonparametric statistical models.

There was a low/moderate concordance between cell phone handedness and the laterality of the tumor in the cerebral cortex (Kappa = 0.26). The Kappa statistic for landline telephones and the laterality of cortical tumors was 0.04. Together, these tests suggest a low/moderate degree of concordance between tumor laterality and cell phone handedness but not for household phone handedness. It is useful to consider the anatomical aspects of cell phone use and RF exposure. It is evident that the facial areas of the ear and cheek on the side of the head where the cell phone is used are exposed to the highest dose of cell phone energy. Because the source exposure does not readily predict biological exposure (1), human models are used to measure the RF rate of energy absorption in terms of specific absorption rate (SAR) which is the rate of total electromagnetic energy absorption in biological tissue. SAR studies show that intracranial absorption of cell phone RF energy dissipates exponentially with distance from the cell phone antenna. The Kappa statistic should be interpreted as an assessment between cell phone handedness and the cerebral hemisphere (right vs. left) of the tumor.

The SAR in the head of a phantom model of a cell phone user was highest in the ear and surrounding brain tissue (2). In humans, this would occur at the cortex in the temporal lobe. If the SARs for the head from cell phone exposure accurately reflect human exposure conditions, then the most appropriate statistical tests for these analysis are based on case patients with temporal cortex tumors. The OR (0.9) and Kappa (-0.24) for the temporal cortex support a lack of an association. Also, the anatomic distribution of brain tumors in this study is similar to other large case series of brain cancers (3).

In the United States, brain cancer rates in the elderly (≥ 65 years) increased 37% in the past 20 years (4). Speculations that this increase is due to analog cell phone use (5) are unfounded since these and other data show that the elderly use cell phones infrequently.

There are several potential sources of bias and problems with exposure measurement in case-control studies of cell phone use. Hospital referral patterns could have biased the results if catchment areas differ between cancer and control patients. Because 93% of cases and 92% of controls lived in New York or New Jersey, referral patterns probably did not affect the findings although there might be more localized referral patterns that went unaccounted. Another possible source of bias is the use of hospital controls rather than community controls as a comparison group. We observed little heterogeneity of cell phone use across different control diagnostic groups and no case-control differences in years of formal education or in types of occupation. Further, there was a similar age and gender distribution between cell phone users in this control group and in the Epidemiology Resources, Inc. (ERI) cohort of more than 250,000 handheld and other mobile phone users in Boston, Chicago, Dallas, and Washington, DC (6). In 1994, the number of handheld cell phone users in male members of the ERI cohort increased with age from 1,904 in the 20-24 age group to approximately 5,000 in each of the age groups 30-34, 35-39, 40-44, and 45-49. There was a marked decline in the number of users that were ≥70 years old. The ratio of male/female handheld cell phone users was 2:1. The control group of the current study had a similar age distribution of handheld cell phones users and the gender ratio of cell phone use was similar (1.7:1).

None of the controls in this study were admitted for injuries resulting from motor vehicle accidents. Motor vehicle accidents were not a diagnostic criterion that excluded subjects from selection into the study. The inclusion of patients admitted for car accidents might have biased the findings since cell phone use has been associated with increased vehicular accidents (7, 8). Problems with recall or memory in both cases and controls is another potential source of bias. In the ERI cohort which obtained actual billing records, the correlation between self-reported frequency of use in mobile cell phone users and the billing records in a 3-month period was 0.74. In this study, there was a moderate correlation ($r = 0.56$) between frequency of use and monthly billing data. The lower correlation found in this study compared to the ERI results might reflect the differences between actual and self-reported billing data, poorer subject health, and averaging both the frequency of use and the bill for the entire period that the cell phone was used. Despite these limitations, the moderate correlation coefficient in this study suggests that self-reported usage information was accurate enough for epidemiologic assessment. Another important exposure variable was information on handedness. In a survey of 3,157 users of handheld and other types of transportable cell phones, handedness was weakly correlated with the side of the head used to hold the cell phone (r = 0.15) (9). Less than 10% of respondents reported switching hands when using the cell phone. We specifically collected information on telephone handedness, and the 9.4% of cell phone users in this study that reported no hand preference provided reassurance that this information was accurate.

It is unlikely that the lack of an association with cell phone use in this study is due to poor recall in cases and controls. Nearly all cell phone users were current users at the time of study entry. Since cases might be expected to attribute their illness to environmental risk factors more often than controls, recall bias would be expected to bias the results toward a positive association (10).

Several factors affect cell phone signal intensity and possibly biological exposure to cell phone energy. Intracranial exposure to cell phone RF energy is reduced with the extension of the antenna. The majority of study subjects reported using cell phones with the antenna extended, with little differences between cases and controls. Information was not obtained on the number of cell phone calls. There is a reduction in signal intensity after the connection is secured and biological exposure might be higher for frequent calls than infrequent calls per calling minute. Although the number of calls as well as whether the user was holding the phone to the ear at the time of connection is difficult to measure accurately, if at all, by questionnaire data, the connection time is established in milliseconds and inconsequential in terms of cancer risk. The amount of power radiating from the cell phone also increases with distance from the base station. The density of cell towers is higher in urban areas than in suburban or rural surroundings. In contrast, power output from the phone increases with cellular traffic which is greater in urban areas. It was not possible to identify the specific location of base station towers or the dates when new towers were constructed during the timeframe of the study. Data on base station locations would likely be uninformative for exposure assessment purposes since most cell phone users reported cell phone usage in an automobile or during commuting. Another determinant of cell phone RF energy exposure is the cell phone model. Power output from older models of cell phones are higher than newer models although the difference might not be meaningful since power output is reduced after the connection is established. Other factors that might interfere with the signal or change the cranial location of peak energy absorption in the head include use of wireframe glasses, metal head implants, or earrings. These factors were not statistical confounders or effect modifiers in this study.

There was no association between the risk of brain cancer and surrogate measures of exposure to home and workplace electric power. Although this information was collected to control for possible statistical confounding, several reviews have concluded that the risk of brain cancer from electric and magnetic fields from power lines are small or hypothetical (11, 12, 13).

In regard to the histologic-specific analysis, there were 2 case patients with ganglioglioma of the brain that received post-operative care at a different participating institution from the surgical center and had their slides reread. The subsequent histologic designation was astrocytoma and glioma, respectively. There have been recent changes in the morphologic criteria for brain cancer (14), and statistical findings based on histologic differences in a multi-center study need to be interpreted with caution. Further studies should seek to replicate these findings.

Risk assessment of cell phone use and brain cancer should take into consideration human and experimental studies of RF exposure. In a review of the limited studies of occupationally exposed workers, it was concluded that there is a small or lack of association between RF energy and brain cancer (15). In a review of experimental studies of RF energy exposure, Brusick et al. (16) concluded that there are limited data to support a genotoxic effect. In a study that tested intermittent digital field exposures for a 2-year period on the incidence of central nervous system tumors in a rat model, a significant reduction of brain tumors occurred in the microwave-exposed group (17). This model study was based on experimental conditions that maximizes the interpolation of the findings to human risk. For instance, the SAR in the experimental animals was set at levels similar to human model conditions. Clearly, there is a need for continued studies of brain cancer risk and longer term usage of cell phones.

Editor's Note

Since the 1999 Colloquium, there has been much controversy in the popular media regarding the significance of this research. The sponsor of the study (G.L.C.) and the author (J.M.) remain in disagreement about these results and their significance to public health.

G.L.C.

References

1. Kuster N & Balzano Q. Energy absorption mechanisms by biological bodies in the near field of dipole antennas above 300 MHz. IEEE Transactions on Vehicular Technology 1992; 41: 174-181.

2. Guy AW. The starting point: Wireless Technology Research, LLC's dosimetry risk evaluation research. Human and Ecological Risk Assessment 1997; 3: 25-50.

3. Preston-Martin S, Pogoda JM, Schlehofer B, Blettner M, Howe GR, Ryan P, et al. An international case-control study of adult glioma and meningioma: The role of head trauma. Inter J Epidemiol 1998; 27: 579-586.

4. Ries LAG, Kosary CL, Hankey BF, Miller BA, Harras A, Edwards BK (editors). SEER cancer statistics review 1973-1994. NIH Pub. No. 97-2789. Bethesda, MD: National Cancer Institute; 1997.

5. Davidson JA. Brain tumors and mobile phones? Med J Aust 1998; 168: 48.

6. Rothman KJ, Loughlin JE, Funch DP, & Dreyer NA. Overall mortality of cellular telephone customers. Epidemiology 1996; 7: 303-305.

7. Redelmeier DA & Tibshrani RJ. Association between cellular telephone calls and motor vehicle collisions. New Eng J Med 1997; 336: 453-458.

8. Violanti JM. Cellular phones and fatal traffic collisions. Accid Anal Prev 1998; 30: 519-524.

9. Funch DP, Rothman KJ, Loughlin JE, & Dreyer NA. Utility of telephone company records for epidemiologic studies of cellular telephones. Epidemiology 1996; 7: 299-302.

10. Wynder EL, Higgins IT, & Harris RE. The wish bias. J Clin Epidemiol 1990; 43: 619-621.

11. Kheifets LI, Afifi AA, Buffler PA, & Zhang ZW. Occupational electric and magnetic field exposure and brain cancer: A meta-analysis. J Occup Environ Med 1995; 37: 1327-1341.

12. Valberg PA. Electric and magnetic fields (EMF): What do we know about the health effects? Int Arch Occup Environ Health 1996; 68: 448-454.

13. National Institute of Environmental Health Sciences (NIEHS). Assessment of health effects from exposure to power-line frequency electric and magnetic fields. Working group report. NIH Pub. No. 98-3981. Research Triangle Park, NC: National Institute of Environmental Health Sciences, US National Institutes of Health; 1998.

14. Joshi VV, Cantor AB, Altschuler G, Cohen LJ, Larkin EW, Shuster JJ, et al. Conventional versus modified morphologic criteria for ganglioneuroblastoma: A review of cases from the Pediatric Oncology Group. Arch Path Lab Med 1996; 120: 859-865.

15. Elwood JM. A critical review of epidemiologic studies of radiofrequency exposure and human cancers. Environ Health Perspect 1999; 107(Suppl 1): 155-168.

16. Brusick D, Albertini R, McRee D, Williams G, Hanawalt P, & Preston J. Genotoxicity of radiofrequency radiation. Environ Mol Mutagen 1998; 32: 1-16.

17. Adey WR, Byus CV, Cain CD, Higgins RJ, Jones RA, Kean CJ, et al. Spontaneous and nitrosourea-induced primary tumors of the central nervous system in Fischer 344 rats chronically exposed to 836 MHz modulated microwaves. Radiat Res 1999 (in press).

22

HUMAN STUDIES ON THE ENDOCRINE AND NERVOUS SYSTEMS UNDER EXPOSURE TO MOBILE PHONES

R. de Seze, L. Miro, & J. Ayoub

In vitro and in vivo experiments cannot always be extrapolated to human health, and epidemiology studies take a long time to obtain results. We therefore decided to perform human studies. These humans studies focused on the endocrine system, the auditory system, and the nervous system. The research was funded by CNET from France Telecom and Motorola.

The studies on the endocrine system looked at blood hormone levels. First we measured baseline hormone levels during the day, second we measured the levels on the circadian chronobiological rhythms, and third we checked whether a stimulated secretion would be more sensitive to electromagnetic fields. The exposure protocol consisted of exposing volunteers for 2 hours per day, 5 days per week, for 4 weeks. To catch the attention of the volunteers and to be sure that they would correctly hold the telephone, voice projection of a 2-hour movie they were watching was fitted to their phone. Blood samples were collected once each week, at 8 AM, 3 weeks before the beginning of the exposure period, at the end of each exposure week, and then for 2 more weeks to look for the retentivity of any eventual effect.

We looked at the pituitary hormones and among them we observed a significant decrease only in the TSH hormone on the last sampling of the exposure period; the decrease was about 21%. We wanted to make sure this effect was not due to a shift in the circadian profile. In a study funded by Motorola we looked at the circadian rhythm of ante-pituitary hormones and also of melatonin which has been shown to be sensitive to extremely low frequency electromagnetic fields in in vivo studies. We observed the hormone profiles of melatonin before the exposure period, after 2 weeks of exposure, after 4 weeks of exposure, and 2 weeks after the end of the exposure period. The profiles were very similar for GSM and DCS phones. In this study, we also observed a significant decrease in growth hormone and in cortisol hormone, but these were slight changes which need to be further examined.

We hypothesized that dynamic secretions would be more sensitive to electromagnetic fields, such as what occurs when there is a pathological or physiological secretion. We performed a stimulated secretion by injecting releasing factors: thyroid releasing hormone (TRH) or gonadostimulin (LH-RH). After TRH injection, we looked at the hormone profiles of thyroid stimulating hormone and prolactin; after LH-RH injection, we looked at the hormone profiles of luteinizing hormone and follicle stimulating hormone. For each of these releasing factors, we performed two tests: the first one during a sham exposure for all volunteers and the second one 15 days later, during an actual exposure for the exposed group, and again during a sham exposure for the sham group. The sham group showed that there is excellent reproducibility of the stimulation test because of a hormone profile very comparable to the exposure group. The same reproducibilty was obtained with the four hormones assayed.

When we looked at the auditory system, nearest to the radiotelephone, we looked at the auditory brainstem responses and the distortion products of otoemission. On the auditory brainstem responses, our interest was in the wave latencies (I, II, III, and V) and the wave intervals (I-III, I-V, III-V). These responses were measured before and after 1 hour of exposure to mobile phones, and we found that before and after exposure, there was no difference of any of the observed parameters. A significant difference in the distortion products of otoemission was not observed either. Experiments were also run on an electroencephalogram (EEG) as some changes were found by Röschke and Thuroczy, although not by Hietanen. At this point, we wanted to find out if the EEGs could physiologically support any potential harmful effect of mobile phones. We studied the effect of a 1-hour exposure on an EEG and measured the power spectral density. Significant differences were observed between the pre-exposure and the post-exposure recordings for different frequency bands and different locations such as right hemisphere, vertex, and left hemisphere. Differences in exposure conditions were also looked at: with a telephone on the ear or without a telephone on the ear. In contrast, after 1-hour sham exposure, there was absolutely no difference in the control group.

In summary, we found no differences from the baseline data in the endocrine system. On the circadian rhythm there was no change in melatonin but a small change in the growth hormone and cortisol; it is thus necessary to reproduce this study. We also did not find any effect on a stimulated secretion. In the auditory system, we did not find an effect on auditory brainstem responses or on distortion products of otoemission, but we observed significant differences on the EEG. Again it is imperative that we check the reproducibility of this change and look at its physiopathology, especially in epileptic patients. Moreover, we need to focus on any possible functional significance within cognitive function, memory, and alertness. These studies are planned to be carried out in the framework of the COMOBIO program launched by the RNRT in France.

23

EFFECTS OF A 915 MHZ SIMULATED MOBILE PHONE TRANSMISSION ON CEREBRAL FUNCTION

A. Preece & J. Eavis

The work being conducted at the University of Bristol in the United Kingdom looks at the effect that mobile phones have on cognitive function. This study was carried out by Dr. Alan Preece and was funded by the Government Department of Health.

In Great Britain mobile phones operate at two different frequency bands, analog and digital. GSM phones operate at about 900 MHz and digital DCS phones operate at about 1800 MHz. At 900 MHz, the penetration depth in tissue is 2 to 3 cm, a value which decreases with increasing frequency. If the specific absorption rate (SAR) values produced by mobile phones are examined, previous work using simulated analog signals shows the maximum SAR values to be in the range of 3 to 8.2 W/kg. More recently, work looking at SAR values in the brain, such as that carried out at CNET in Italy, has shown a value of 0.3 W/kg. In the UK, the National Physical Laboratory is undertaking SAR measurements of real mobile phones and has shown values between 0.02 and 0.44 W/kg, where the main variation is in the design of the handset and, in particular, the position of the aerial. To briefly summarize, the highest brain SAR values from digital GSM phones have been measured at about 0.44 W/kg and for analog phones 0.9 to 1.76 W/kg.

In our study we have been looking at the effects mobile phones have on cognitive function. Cognitive function tests are routinely used to assess the effects that drugs have. This work used a series of 10 different tests looking at speed, accuracy, attention, and memory. We measured reaction times, word, number, and picture recall, and spatial memory.

The work used two groups of volunteers, each group containing 18 members, and each member was asked to perform the 10 cognitive function tests while being exposed to three different exposure regimes. The first exposure was a simulated analog signal of 1 W continuous wave 915 MHz. The second exposure was a simulated digital GSM signal where the same continuous wave signal was modulated at 217 Hz with a 12.5% duty cycle. The third exposure was a control exposure whereby the radiofrequency power was delivered from the phone itself and into a 50-Ω dammy load. In each case, the user wore a mock mobile phone positioned on an ear defender so that the antenna was above the left squamous temporal bone. The cognitive function tests were performed on a standard PC using a high-definition monitor. Only the choice reaction time test was statistically significant. This test works with either a 'yes' or 'no' being flashed on the PC screen. The subject has a control panel with a 'yes' button and a 'no' button, and the subject has to press the corresponding button in the shortest possible time. None of the other parameters were statistically significant.

An alternative analysis performed was the multivariate analysis whereby the 15 data sets produced from the original 10 cognitive function tests were pooled into four different categories. Again, the only one with a P-value that showed any statistic significance was the reaction time on attentional tasks.

In summary, we believe we have shown that there is an effect on the brain at 915 MHz transmissions. The effect is larger for analog signals than it is for digital GSM transmissions. The main effect is a 15-millisecond reduction in the transit time between the visual and speech senses in the brain. Because of the length of this reduction, we believe it is a synaptic effect rather than an axon transmission effect. The question remains as to whether the overall effect is a thermal effect or a nonthermal effect.

24

SYMPTOMS EXPERIENCED IN CONNECTION WITH THE USE OF DIGITAL AND ANALOG MOBILE PHONES FOR PEOPLE USING BOTH TRANSMITTER SYSTEMS: A SWEDISH-NORWEGIAN EPIDEMIOLOGICAL STUDY

G. Oftedal, M. Sandström, K. Hansson Mild, & J. Wilén

Introduction

The two most commonly used transmitter systems in Sweden and Norway are, and were at the time of this study, the digital GSM system and the analog NMT system, both systems operating at frequencies around 900 MHz. During 1995 many Norwegian and Swedish consumers contacted us with complaints of headaches, discomfort, warmth sensation, and other symptoms when using the mobile phone (MP). The following pattern seemed to emerge: More GSM users than NMT users complained and some had experienced the symptoms when they changed from the old NMT phone to the new digital GSM phone.

We initiated a Swedish-Norwegian epidemiological study to test whether the prevalence of symptoms was higher among GSM users than among NMT users. Adjusted odds ratios did not indicate any increased risk for symptoms among GSM users compared to NMT users. However, sensations of warmth on the ear and around/behind the ear were more often experienced by NMT users than by GSM users in both countries with statistical significance. Similar results were obtained for headaches in Sweden (1). These results were obtained from questions about symptoms experienced in general, that is, the symptoms were not attributed to any specific device. In this paper we have focused on symptoms attributed to MP use and the following question: Do people using both a GSM phone and a NMT phone experience symptoms more often in connection with the use of the GSM phone than in connection with the use of the NMT phone or visa versa?

Methods

Questionnaires were sent to people using a MP in their job and to approximately the same number of people with a GSM subscription as with a NMT subscription—5,000 total in Norway and about 12,000 in Sweden. The response rate adjusted for people not receiving the questionnaire was 64% in Norway and 76% in Sweden. The number of respondents using both a GSM phone and a NMT phone was 484 in Norway and 1,152 in Sweden.

In the questionnaire we asked questions about symptoms, exposure factors concerning the MP, the use of the MP, and potential confounding factors. We asked both how often the symptoms were experienced (without attributing to any device), and whether the symptoms were experienced in connection with MP use. For more details see Oftedal et al. (1).

Binomial distribution (sign test) was used to test whether the number of people that had attributed a symptom to the GSM phone was different from the number that had attributed the symptom to the NMT phone. A result was considered to be statistically significant if the significance probability was less than 0.05.

Results

Out of the respondents in both Norway and Sweden using both a NMT and a GSM MP, people generally had used the NMT phone for a longer period of time than the GSM phone. In Norway the number of calls per day and the total calling time per day was approximately equal for the two MPs. Slightly more calls per day and longer calling times per day were observed for the GSM phone than for the NMT phone in Sweden (see Table 1).

Respondents using both a NMT phone and a GSM phone most often attributed the symptoms to the GSM phone. Statistically significant differences were obtained for a few symptoms in Sweden. For these symptoms the Norwegian results are more or less in accordance with the Swedish; differences are not statistically significant. In Norway memory loss was statistically significantly more often attributed to the NMT phone than to the GSM phone, whereas in Sweden this symptom was most often (but not statistically significantly) attributed to the GSM phone. Many people attributed the symptoms to both the GSM and the NMT MP (see Table 2). To compare the results for users of both transmitter systems with those obtained when comparing symptom prevalence among GSM users and NMT users, we calculated the ratio between the percentage having a GSM-attributed symptom and the percentage having a NMT-attributed symptom. Furthermore, we have calculated the ratio between the prevalence of symptoms among the GSM users and the NMT users. A value higher than one (see Figure 1) is obtained when the number of people with a GSM-attributed symptom is highest, that is, the prevalence of symptoms is highest among the GSM users.

In Norway the difference between the results obtained with the two methods is small for most symptoms. In Sweden people using both a GSM and a NMT phone

Factor	Category	Norway (n=484)		Sweden (n=1152)	
		NMT (%)	GSM (%)	NMT (%)	GSM (%)
Time with MP	< 13	47	74	31	60
	13 – 24	18	16	20	25
	> 24	29	3.9	41	11
	Missing	5.2	6.4	7.6	4.5
No. of calls/day	< 2	20	23	27	26
	2 – 4	21	20	22	23
	> 4	55	55	39	48
	Missing	3.5	2.1	11	3.3
Calling time/day	< 2	13	15	23	20
	2 – 15	40	40	38	41
	15 – 60	31	34	23	30
	> 60	11	9.3	4.1	5.4
	Missing	4.5	2.1	11	3.4

Table 1. The distribution of respondents using both a NMT and a GSM MP with respect to the period of time with the MP, the number of calls per day, and the total calling time per day for the NMT phone and the GSM phone, respectively, and the number of people (n) for Norway and Sweden, respectively.

Symptom	Norway (n=484)				Sweden (n=1152)			
	Both	NMT	GSM	p	Both	NMT	GSM	p
Dizziness	12	16	17	1.00	5	5	10	0.30
Discomfort	8	11	12	1.00	8	4	22	<0.001
Concentration	9	10	9	1.00	6	8	9	1.00
Memory loss	6	12	2	0.01	5	2	5	0.45
Fatigue	9	15	18	0.73	6	4	10	0.18
Headaches	24	16	21	0.51	14	9	16	0.23
Warmth around/behind ear	56	39	29	0.27	34	30	44	0.13
Warmth on ear	61	40	35	0.64	52	41	41	1.00
Burning skin	21	14	20	0.39	14	19	16	0.74
Tingling/tightness	7	7	16	0.09	7	5	15	0.04
Other	5	5	7	0.77	7	6	18	0.02

Table 2. The number of people who attributed the symptom to both a NMT and a GSM phone (Both), to a NMT phone (NMT), and to a GSM phone (GSM), and the significance probability (p) (binomial distribution) of a difference in the number with a NMT-attributed symptom and with a GSM-attributed symptom. Respondents (n) in Norway and Sweden, respectively, with both a NMT and a GSM phone are included.

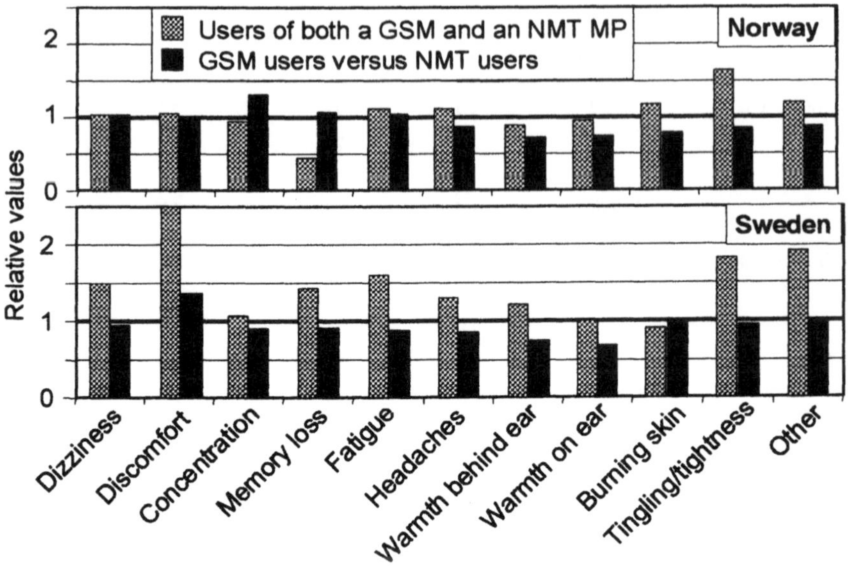

Figure 1. The ratio between the number of respondents that attributed symptoms to a GSM phone and the number that attributed symptoms to a NMT phone for people using both transmitter system (grey bars) and the ratio between prevalence of symptoms between GSM users and NMT users (black bars).

tended to have relatively more symptoms attributed to the GSM phone than would be expected by symptoms among GSM users compared to NMT users.

Discussion

If the GSM phone causes more symptoms than the NMT phones, we would expect similar results in both Sweden and Norway. We would also expect consistency between results obtained for people using both transmitter systems and results obtained when GSM users are compared to NMT users. Confounding may be one reason for the observed deviations. The results for the GSM users versus NMT users in Figure 1 are based on crude prevalence data. Adjustment for potential confounding factors, however, does not change these results much.

Among all respondents we have observed that the number of people with any symptom attributed to MP use increases with the number of calls per day and with the total calling time per day. Since the Swedish respondents with both transmitter systems used their GSM most (Table 1), this may explain to some extent why these respondents more often experienced symptoms in connection with use of the GSM phone than in connection with use of the NMT phone.

Among other possible reasons for a dominance of GSM-attributed symptoms is a recall bias. The GSM system was introduced just a couple of years before this study, and some people might have changed from a NMT phone to a GSM phone so that they used only the GSM phone at the time they filled in the questionnaire. It is likely that people would remember recent symptoms better than symptoms experienced some months ago.

We should also be aware that the number of people having attributed a symptom to either the GSM phone or to the NMT phone was relatively low for most symptoms. Therefore, we should take this uncertainty into account when evaluating these results.

Conclusion

In this study, people using both a NMT phone and a GSM phone attributed the symptoms more often to the GSM phone than to the NMT phone, but statistically significance was reached for only some symptoms and for only Sweden. The results might have been confounded to some extent and also influenced by recall bias. Therefore, it is not sufficient evidence to conclude that there is a difference in symptoms experienced in connection with the use of GSM phones compared to the use of NMT phones for people using both a GSM and a NMT phone.

Acknowledgment

We received financial support for the study from The Swedish Mobil Telecommunications Association (MTB), The Work Environment Fund of the Confederation of Norwegian Business and Industry, Telenor, The Norwegian Post, and Telecommunication Authority.

Reference

1. Oftedal G, Hansson Mild K, Sandström M, Wilén J, Tynes T, Haugsdal B, & Hauger E. Comparison of symptoms experienced by users of analogue and digital mobile phones: A Swedish-Norwegian epidemiological study. SINTEF Report STF78 A98134. Trondheim:, 1998.

VI

FUTURE DIRECTIONS

25

SURVEILLANCE AND RESEARCH NEEDS: THE NEXT FIVE YEARS

G.L. Carlo

It is important within the topic of wireless phones and health to discuss recommendations and future directions, particularly when you have the opportunity during a colloquium gathering scientists from around the world. It is also very important at this time as it is probably the last time I will be speaking as Chairman of Wireless Technology Research, LLC (WTR) and perhaps the last time there is a WTR activity. It is imperative to me and to those who have participated in the WTR program to make sure that our thoughts and our concerns about the future are properly conveyed. In this way, those that will be charged with taking up the activities from this point forward will have the benefit of all of our knowledge as well as all of our mistakes.

There is a new wireless customer every 2.25 seconds. There are more than 500 million people counting on us to get this right. This is a very rapidly growing and changing technology. Digital subscribership has increased 183.5% in the last year. From a public health point of view, we are trying to simultaneously accomplish surveillance, tracking, monitoring, and the ability to respond quickly so the consumers can be protected. At the same time we have to conduct research that is rigorous, is peer reviewed, and will stand the test of time. All of this put together is a very difficult task. Epidemiological studies and definitive animal studies take years to complete which limits their value to rapidly changing technologies. Epidemiology studies also have disease latency limitations even with static technologies. Wireless technology is not only rapidly growing but rapidly evolving. The analog, digital, TDMA, CDMA, and PCS technologies and the profile of the penetration of those technologies around the world is changing every day. The handsets themselves and the configuration of the antennas are changing very rapidly and are presenting a problem, the likes of which we have really never seen before.

Over the past 6 years, WTR has had the opportunity to do both the surveillance and research in the United States. We have learned a great deal about the limitations of conventional approaches to both surveillance and research. We have done the best we could to lay a foundation so that the short-term and long-term melding of surveillance, tracking, monitoring, and research could be accomplished. With all that taken into account, we have some specific recommendations that we would like to leave with you. First, there needs to be technology surveillance and passive surveillance or a passive complaint reporting system. There also needs to be active public health surveillance through follow-up of three specific cohorts. I will forego discussing research recommendations and instead defer to Dr. Repacholi and his team at the World Health Organization and the International Agency for Research on Cancer who are doing a magnificent job defining the research needs that will address the types of problems that are emerging. Instead, I will address the needs that relate to consumer protection, the needs arising from generated data filtered through a public health paradigm for the protection of the public. Because the technology is rapidly evolving, it is very important that the baseline information we have now from in vitro genotoxicology tests, in vivo studies, and human epidemiology studies be properly catalogued for future studies. We need to have a program where the technology is monitored as it evolves. Whether it is an evolution of handset technology or antennas, it needs to be monitored and it needs to be periodically tested to pick up any change in technology that will render the existing interpretations, with regards to the baseline data, inadequate. If we do find changes in technology that alter our current conclusions, then we need to reintroduce in vitro studies of short-term measure. We can then compare the results of these new technologies with the results from the existing technologies. This is the only way to really account for the rapidly changing technology.

In regards to the passive reporting system that was mentioned earlier, there needs to be a system in place which captures complaints from wireless phone users. These data need to be maintained in a database for future use and periodic evaluation against a public health paradigm. Within the active public health surveillance system, there are three epidemiological cohorts which should be established and followed through record-linkage methods. These are the cellular analog cohort, the cellular digital cohort with TDMA and CDMA technologies, and a PCS digital cohort with TDMA and CDMA technologies. These cohorts can be established with warranty cards or user records. In the United States and around the world, there are numerous tumor reporting systems, for example, the National Death Index in the U.S. I believe that record-linkage work is probably the most efficient manner in which these large cohorts can be evaluated for early suggestions of potential problems. As you can see, I am not anticipating an end to this any time soon; that is the way public health and surveillance work. As Dr. Elizabeth Jacobson of the US Food & Drug Administration has said on numerous occasions, post-market surveillance, tracking, and monitoring need to be regarded by the wireless industry as an ongoing cost of doing business. Nothing less than that will protect the consumers.

Based on what we have learned since 1993, I believe that we need an immediate and comprehensive review of the 1.6 W/kg radiofrequency emission standard for handheld wireless instruments. The rationale for this is straightforward. First, the current standard is based on historical data only indirectly relevant to the human exposure situation. In 1992, cellular phones were not even included in standards because of a low-powered solution. They were added to the standard following that but with no new data, so that the current standard is a retro-fit. Because historical data are oriented toward thermal effects, it is important to know that if there are any adverse effects from wireless technology, they are not going to be thermal. In the past 5 to 6 years there have been new exposure systems developed both in vitro and in vivo by WTR and by other groups that give us the opportunity to develop directly relevant information. There are exposure systems now that have been engineered to assimilate and approximate exposures from wireless phones. Some of the exposure systems are even driven by wireless phones. Overall, if we have the opportunity to use directly relevant data, wc should. In order to begin this process, I highly recommend that an international regulatory task force become involved. Clearly, wireless communication is global, and research needs be coordinated on a global basis. Consumers are the same around the world, so it makes no sense to have a different standard in different countries.

Finally, driving the ship of WTR has caused us to hit a couple of waves. While there has been some acrimony from a number of people, ranging from those in industry and government to scientists and consumers, one thing has become very clear. While many times I have gone head to head with the Cellular Telecommunications Industry Association (CTIA), it is very clear to me CTIA was doing its job, and I was doing my job. At the same time, while I have had arguments with people from Motorola about exposure systems and about the way data were interpreted, it is clear to me that they were trying to do their job, and I was trying to do mine. As we discuss all of these scientific interplays on this subject over the past 5 years, the scientists were doing their job. The difficulty, however, lies in this: Even though it is clear to me that we are all doing our jobs, it is also clear to me that the consumers are not being protected, which says there is something missing in our current system. As I said earlier, the industry is responding to consumer demand and consumers want this technology. Does it not make sense that if the consumers want this technology they should absorb some of that risk associated with this technology? Does it not make sense that consumers should be empowered to protect themselves? I do not have the answers, but these are questions that will have to be answered before this controversy about wireless technology and public health goes away.

APPENDIX A

SECOND STATE OF THE SCIENCE COLLOQUIUM AGENDA

sponsored by
Wireless Technology Research, LLC

HYATT REGENCY, LONG BEACH, CALIFORNIA, USA

19-20 JUNE 1999

Saturday, 19 June 1999

8:00-9:00 AM	Check in and on-site registration	
9:00- 9:15 AM	**INTRODUCTION TO COLLOQUIUM**	Dr. George L. Carlo Wireless Technology Research, LLC
9:15-10:30 AM	**SESSION I: DOSIMETRY AND MEASUREMENTS** **Chair: Dr. Arthur Guy**	
	State of the Science: RF Dosimetry and Measurements	Dr. Niels Kuster ETH Zurich
	Dosimetry Studies on In Vitro Preparations Exposed to 837 and 1900 MHz in a TEM Cell	Dr. Arthur W. Guy University of Washington
	WTR In Vivo Exposure System	Dr. Donald McRee Wireless Technology Research, LLC
	A Fast Tour of a Decade of Bioelectromagnetic Hygiene Research	Dr. Marjorie Lundquist Bioelectromagnetic Hygiene Institute
10:30-10:45 AM	Break	
10:45-11:15 AM	Floor Statements and Discussion	
11:15-12:15 PM	**SESSION II: CERTIFICATION** **Chair: Dr. Robert Cleveland**	
	Developing a Regulatory Compliance Regime for Electromagnetic Radiation from Radio Transmitters Based on Health Safety Considerations in Australia	Mr. Robert J. Johnston Australian Communications Authority
	Guidelines for Radiofrequency Exposure from Mobile and Portable Devices Adopted by the US Federal Communications Commission	Dr. Robert Cleveland Federal Communications Commission
	FCC Equipment Authorization Procedures for RF Exposure Compliance of Mobile and Portable Transmitters	Mr. Kwok Chan Federal Communications Commission
12:15-12:45 PM	Floor Statements and Discussion	

12:45-1:45 PM	Lunch (on your own)	
1:45-3:30 PM	**SESSION III: TOXICOLOGY** ***In Vitro Studies/Genotoxicology*** **Chair: Dr. Joseph Elder**	
	State of the Science: RF Genetic Toxicology	Dr. Joseph Morrissey Motorola
	Evaluation of Potential Genotoxicity of Radiofrequency Radiation Technologies Using a Battery of In Vitro Tests	Dr. Graham Hook Integrated Laboratory Systems
	Investigation of DNA and Protein/DNA Crosslinks in Murine C3H/10T1/2 Fibroblasts after Exposure to 2450 MHz Electromagnetic Radiation	Dr. Isabelle Lagroye University of Bordeaux
	Evaluation of Potential Genotoxicity of Radiofrequency Radiation Technologies Using the In Vitro Single Cell Gel Electrophoresis and Micronucleus Assays	Dr. Graham Hook Integrated Laboratory Systems
	Micronucleus Assay: A Summary of Scientific Discussion	Dr. Donald McRee Wireless Technology Research, LLC
3:30-3:45 PM	Break	
3:45-4:30 PM	Floor Statements and Discussion	
4:30-4:45 PM	**CLOSING REMARKS AND INTRODUCTION TO DAY TWO**	Dr. George L. Carlo Wireless Technology Research, LLC
5:00-6:30 PM	Reception (Harbour Room)	

Sunday, 20 June 1999

9:00-9:15 AM	**OPENING REMARKS**	Dr. George L. Carlo Wireless Technology Research, LLC
9:15-11:00 AM	**SESSION IV: TOXICOLOGY** ***In Vivo Studies*** **Chair: Dr. Michael Repacholi**	
	State of the Science: In Vivo Toxicology	Dr. Michael Repacholi World Health Organization
	Ongoing Studies on the Biological Effects of 835.62 MHz FMCW and 847.74 MHz CDMA Radiation	Dr. Joseph Roti Roti Washington University Medical Center
	Effects of GSM-900 Microwaves on Rat Models	Dr. Rachid Anane University of Bordeaux

	Evaluation of Potential Genotoxicity of Radiofrequency Radiation Technologies Using the In Vivo Single Cell Gel Electrophoresis (SCGE) Assay: Integrated Laboratory Systems Study	Dr. Raymond Tice Integrated Laboratory Systems
	Evaluation of Potential Genotoxicity of Radiofrequency Radiation Technologies Using the In Vivo Single Cell Gel Electrophoresis (SCGE) Assay: University of Washington Study; Comparison of ILS and UW SCGE Studies	Dr. Graham J. Hook Integrated Laboratory Systems
11:00-11:15 AM	Break	
11:15-12:00 PM	Floor Statements and Discussion	
12:00-1:00 PM	Lunch (on your own)	
1:00-3:00 PM	**SESSION V: EPIDEMIOLOGY** **Chair: Dr. George L. Carlo**	
	State of the Science: RF Epidemiology	Dr. George L. Carlo Wireless Technology Research, LLC
	Use of Cellular Telephones and the Risk for Brain Tumors: A Case-Control Study	Dr. Kjell Mild National Institute for Working Life
	WHO's Key RF Research Requirements for Better Health Risk Assessments	Dr. Michael Repacholi World Health Organization
	Epidemiologic Study of Cellular Telephone Use and Malignant Brain Tumors and Acoustic Neuroma	Mr. Joshua Muscat American Health Foundation
3:00-3:15 PM	Break	
3:15-4:00 PM	Floor Statements and Discussion	
	Invited Statement	Dr. Rene de Seze Laboratoire de Biophysique Medicale-Faculte de Medecine
	Invited Statement	Mr. Joe Eavis Medical Physics University Research Center
4:00-4:45 PM	**SURVEILLANCE AND RESEARCH NEEDS: THE NEXT FIVE YEARS**	Dr. George L. Carlo Wireless Technology Research, LLC
4:45-5:30 PM	Discussion and Closing	

APPENDIX B

COLLOQUIUM PARTICIPANTS

Dr. George L. Carlo
Chairman
Wireless Technology Research, LLC
1711 N Street, NW
Suite 400
Washington, DC 20036-2811
United States

Dosimetry and Measurements

Dr. Niels Kuster
Professor
ETH Zurich
Laboratory for EMF and Microwave Electronics
Zurich
CH-8092
Switzerland

Dr. Arthur W. Guy (Session Chair)
Bioelectromagnetics Consulting
18122 Sixtieth Place, NE
Kenmore, WA 98028
United States

Dr. Donald McRee
Director of Extramural Research
Wireless Technology Research, LLC
2602 Cooleemee Drive
Raleigh, NC 27608
United States

Dr. Marjorie A. Lundquist
Bioelectromagnetic Hygienist
Bioelectromagnetic Hygiene Institute
P.O. Box 11831
Milwaukee, WI 53211-0831
United States

Certification

Mr. Robert J. Johnston
Manager, EMR Effects
Australian Communication Authority
Standards & Compliance Group
12/200 Queens Street
Melbourne, Victoria 3000
Australia

Dr. Robert Cleveland (Session Chair)
Physical Scientist
Federal Communications Commission
Office of Engineering and Technology
2000 M Street, NW, Room 230
Washington, DC 20554
United States

Mr. Kwok Chan
Federal Communications Commission
7435 Oakland Mills Road
Columbia, MD 21046
United States

Toxicology: In Vitro Studies/Genotoxicology

Dr. Joseph Elder (Session Chair)
US Environmental Protection Agency
National Health and Environmental Effects Research Laboratory
MD-85
Research Triangle Park, NC 27711
United States

Dr. Joseph Morrissey
Motorola, Inc.
Corporate Research Laboratory
8000 West Sunrise Boulevard
Fort Lauderdale, FL 33322
United States

Dr. Graham J. Hook
Integrated Laboratory Systems
P.O. Box 13501
Research Triangle Park, NC 27709
United States

Dr. Isabelle Lagroye
Assistant Professor
EPHE
PIOM
BP 108, Avenue Pey Berland
Talence 33402
France

Dr. Donald McRee
Director of Extramural Research
Wireless Technology Research, LLC
2602 Cooleemee Drive
Raleigh, NC 27608
United States

Toxicology: In Vivo Studies

Dr. Michael H. Repacholi (Session Chair)
Scientist
World Health Organization
EHG/EHR
Geneva 27
CH-1211
Switerzerland

Dr. Joseph Roti Roti
Washington University Medical Center
Mallinckrodt Institute of Radiology
4511 Forest Park Boulevard
Saint Louis, MO 63108
United States

Dr. Rachid Anane
Post-Doctural Researcher
PIOM Laboratory
ENSCPB
BP 108
Talence 33402
France

Dr. Raymond Tice
Vice President, Research & Development
Integrated Laboratory Systems
P.O. Box 13501
Research Triangle Park, NC 27709
United States

Dr. Graham J. Hook
Integrated Laboratory Systems
P.O. Box 13501
Research Triangle Park, NC 27709
United States

Epidemiology

Dr. George L. Carlo (Session Chair)
Chairman
Wireless Technology Research, LLC
1711 N Street, NW
Suite 400
Washington, DC 20036-2811
United States

Dr. Kjell Hansson Mild
National Institute for Working Life
P.O. Box 7654
Umea S-907 13
Sweden

Dr. Michael H. Repacholi
Scientist
World Health Organization
EHG/EHR
Geneva 27
CH-1211
Switerzerland

Mr. Joshua Muscat
American Health Foundation
Epidemiology
320 East Forty-third Street
Third Floor
New York, NY 10017
United States

Dr. Rene de Seze
CHU Nimes
Biophysique Medicale
Faculte De Medecine
Nimes, Cedex 2
30 907
France

Mr. Joe Eavis
Research Associate
University of Bristol
Medical Physics Research Centre
Bristol Oncology Center
Horfield Road
Bristol BS2 8ED
United Kingdom

APPENDIX C-1

DOSIMETRY AND MEASUREMENTS

Dr. Guglielmo D'Inzeo, University of Rome:
I have one question related to Dr. Lundquist's transparencies where you showed facts on the blood-brain barrier from a 1977 paper. I have one comment and one question on that. There are several papers on the blood-brain barrier, and you extrapolate some results from just one paper. That normally is not so efficient for protection reasons. Also, there are frequency windows, not just amplitude windows. Do you know the frequency of this experiment and how it is related to mobile phone frequencies?

Dr. Marjorie Lundquist, Bioelectromagnetic Hygiene Institute:
I do not think I can give you the answer you are looking for with respect to that particular paper. What had struck me when I read that paper was that we had this plot that indicated a peaked type of curve that falls off at either end. I showed it here today to indicate that at least one person has gotten experimental results that indicate or suggest that this kind of phenomenon does exist. I am not going to try to defend blood-brain barrier literature because I have not studied it thoroughly enough to know. But here is one group of researchers who published this type of graph and I simply wanted to show that there was some experimental indications of this type of phenomenon.

Dr. D'Inzeo:
There are several papers showing windows in amplitude, frequency, and so on. That is the problem. That is exactly the reason we cannot fix some exact sort of level as safe or as unsafe as you showed before. I think that is a mistake in trying to communicate the risk about these problems.

Dr. Lundquist:

You are right that the power frequency windows and where they are located are important. When I was working for that contractor who had the NIOSH criteria document on microwave radiation contract, the summary of evidence that was made at that time, about 1980, did have a nice little diagram locating these things. I was young and naive and did not realize that I ought to steal my own copy of the stuff at the time so I turned it all back to that contractor as I was supposed to. As a consequence I do not have any of that; I only have my vague memories of it. I have tried through the Freedom of Information Act to try to get copies of this from NIOSH. I have not been successful. Apparently they say that it does not now exist in the form it was then, which means they trashed the stuff they got from the contractor, I guess. The point is the taxpayers of this country paid for a study of the world literature and nice little plots and maps that tell us this kind of thing and would be very useful information, and then the government just sat on it, never published it, never released it. I cannot get it from them now and I do not know whether any body else could. Yes, this is valuable information. So far I do not have it.

Dr. D'Inzeo:

My second question is related to your transparency number 37. You said in mobile phones the dramatic aspects are higher; the dramatic aspects of fields closer to an antenna are higher. What do you mean, the dramatic aspects are higher?

Dr. Lundquist:

In the intermediate or fresnel region around a finite antenna you have rapid fluctuations, rapid in the sense of as you move from point to point in space, fluctuations in the intensity, and they are rather strong fluctuations. And I am saying that the equations indicating that there is a term that depends on the spatial variability of the fields, it would indicate that the fresnel region should then have a much higher contribution to that term and therefore would appear to be reaching a higher hazard. You do not have this in the front of the region, and you do not really have it in the rear region either; that is why I put the fresnel region as the highest risk region on the spaces.

Dr. D'Inzeo:

What you are saying is that the fields there are changing in a quicker way than in another region?

Dr. Lundquist:

Our spatial variation is greater there.

Dr. D'Inzeo:

Why do you say this is dangerous? On what basis?

Dr. Lundquist:

The equation for the volume density of the Lorentz force has the term in the divergence of T, and that is the basis.

Dr. D'Inzeo:
You are speaking of the fields outside the body. Inside the body you have to evaluate the electric fields in the body.

Dr. Lundquist:
It is easier to think conceptually of it in terms of the fields outside the body, but I did provide a form of the equations that is valid inside the body as you are able to measure the fields inside the body.

Dr. D'Inzeo:
My last comment is this: Is it not true that using a digital phone you can have a wider distribution of frequency with respect to an analog one? It seems the bandwidth of the two systems is more or less the same. The bandwidth of the signals is the same, otherwise you can get interference.

Dr. Joseph Morrissey, Motorola:
You mentioned in your talk the connection between brain tumors and cell phones. What evidence to have for this, especially the regional brain tumors you mentioned, in the region of the cell phone?

Dr. Lundquist:
There have been a lot of anecdotal reports and I consider anecdotal reports to be sentinel evidence. In other words, it alerts you to look out for this. The analysis that I offered on the basis of the theory would have indicated that you would have a region of high risk for nonthermal health effects in the brain behind the hair and coupled with the anecdotal reports—at the time the analog cellular phones were the only ones on the market—that people were getting brain tumors in this area is the basis for what I say. Obviously we need good sound, solid epidemiological studies to verify this sort of thing. Whether we will hear about this today I do not know. Sooner or later I think they will appear.

Dr. Morrissey:
What were the anecdotal reports?

Dr. Lundquist:
They are stories I picked up from reading the news media and things on the Internet.

Dr. Morrissey:
Can you give an example?

Dr. Lundquist:
There was a politician—I forget his name—he was in the Republican party who was a heavy user of his cellular phone, and I believe developed just such a tumor and died of it. There have been other similar reports. I have not made a list of names. I have tried in a few cases to get further information. For example, Gene Siskel died not too long ago and I tried to follow up and get some further information on that. Very often

it is difficult to do because the family members or others who are in a position to have access to this information do not want to be disturbed, or it is not that important to them and they do not respond. Trying to get good sound, solid information from anecdotal reports is difficult.

Dr. Joseph Roti Roti, Washington University:
My comment is on the notion that if heat is used to kill cancer cells, it is therefore not carcinogenic. Ionizing radiation and chemotherapeutic agents kill far more cancer cells in clinical practice than heat does, although heat is used in clinical practice. Yet they are clearly carcinogenic and genotoxic agents. The fact is that the data that have been published using transformation assays with heat show that heat is not in fact carcinogenic in and of itself, probably at least by a transformation assay. It has nothing to do with whether or not it kills cells; it has to do with the mechanism by which it kills cells. That is point number one. Number two, heat with something like non-ionizing radiation will increase the carcinogenic frequency, so that it can add to other agents. You cannot therefore just make a blanket statement that because something is used to kill cancer cells clinically, it is not carcinogenic. It depends on how cells are killed and what the target for cell killing is.

Dr. Lundquist:
When one makes arguments, one makes them on the basis of plausibility. They are not confirmed proof. I did not of course cite the experimental study that recently was published which suggests that maybe my conclusion was right even if my argument was flawed. The thing that I wanted to comment on in response to what you said is that first of all nothing in the theory that I am dealing with here is peculiar to non-ionizing radiation. I have just been applying it there because this is the concern of this audience. The thing I have noticed in studying the effects of electromagnetic radiation is that you can get opposing effects under different conditions of exposure to the same agent. As I said to somebody privately here, electromagnetic radiation is an extemely complex agent: It can play games with us and leave us looking like fools. To say that one thing occurs and the opposite cannot is something that can only be said with considerable qualification. It seems the exposure conditions are different to produce one effect and its opposite, and I do not think we understand these differences. I think intensity is one of them. It may not be the only one; I do not know. But I understand what you are saying and I think you are quite right.

APPENDIX C-2

CERTIFICATION

Dr. Andrew Wood, Swinburne University of Technology:
I have a question for Dr. Chan that is in relation to SAR-based testing methodologies. I noticed on your overhead the dielectric constant and the conductivity. I was part of the Australian task group involved in this. We found quite a large variation in sigma values quoted around the world at 800-900 MHz. I am wondering, did you decide that a certain set of parameters were correct?

Dr. Kwok Chan, US Federal Communications Commission:
At this stage the FCC is working closely with the IEEE SCC34 Subcommittee 2 in terms of developing parameters for those. The working groups of that committee have adopted certain basic criteria which are equivalent to some of the references we have on our website, including those of Dr. Camelia Gabriel. Again those are not requirements; essentially we establish a minimum. Generally all the reports we have seen from manufacturers are higher than that. Because of some uncertainties, they choose to provide an overestimation to cover their bases. That is pretty much the trend right now.

Dr. Robert Johnston, Australian Communications Authority:
Basically in Australia we are using the same numbers as the FCC at the moment, based on Camelia Gabriel's work.

Dr. Robert Cleveland, US Federal Communications Commission:
You can look at those values on the FCC website, as Kwok mentioned.

Dr. Wood:
It is just that the sigma value is used in SAR, of course, and if a manufacturer uses a very low value, then a marginal device could pass the test.

Dr. Cleveland:
This is one of the reasons we are working with this committee because we have a lot of experts on this committee and we are trying to come to some consensus on what values to use.

Dr. Chan:
Generally in terms of evaluating compliance, we take into consideration many of the test barometers. Signal is one of them. There are also the test procedures, the final SAR, and certain other things. In certain situations when the SAR is at a certain level, we have to go back and ask for more testing to demonstrate the point.

Dr. Ross Adey, University of California at Riverside:
I have some questions for Dr. Chan. They relate to the techniques and technologies that have evolved to describe the digital phone signal and how this affects the measure of radiated power and the SAR. Briefly there are three parameters. There is the slot average power which for the North American digital phone is the power in one packet and the packet has a 30% duty cycle. But in addition there is a peak power value within the packet which is substantially higher than the slot average power. Then there is a third parameter, time average. Now if we look at the Iridium cell phone, the satellite phone, where there are only 11 packets per second, the duty cycle is substantially reduced. The peak power is in excess of 7 W. One must of necessity from a biomedical point of view consider the peak power as one of the more significant parameters in the tissue interaction. What is the FCC doing to envisage this process rather than, as I assume, using merely a time average measurement?

Dr. Chan:
In terms of measurement procedures for the duty factors and peak powers which you just mentioned, generally we do go back and ask applicants to verify and demonstrate their probe calibrations to ensure that the response of the probe is correct in terms of peak-to-average ratio and certain other things, since in general for RF exposure purposes we are looking at the net absorbed power in the tissue. We have been addressing those things, and the applicants have been required to demonstrate that the calibration of the probe has to work, along with the modulation, the technology, and other things for the transmitter they have submitted.

Dr. Adey:
I take it from one of your last remarks that the energy absorbed in the tissue which can be considered merely the thermalizing dose is what you are using and you are not giving any consideration to any of the modulation parameters which have also well demonstrated to determine the tissue interaction.

Dr. Chan:
The FCC is an agency for enforcing limits that have been adopted, based on recommendations from other government agencies, standards organizations, and many other things. When there are certain concerns, we do bring in the FDA and other government agencies to assess the situation. If there are needs for changes of

limits and certain specific things, it will not just be the FCC; it will be multiple agencies.

Dr. Cleveland:
At this point we are not considering any modulation effects because we are not really the ones to determine that we should be doing that.

APPENDIX C-3

IN VITRO STUDIES/GENOTOXICOLOGY

Dr. Marjorie Lundquist, Bioelectromagnetic Hygiene Institute:

Dr. Morrissey, I do appreciate your point about positive effects needing to be replicated before they can be accepted as scientifically validated. On the other hand, those of us in the health professions know that very often we have to make decisions based on less than ideal information, and we cannot afford to neglect the information at hand just because it does not come to us in the ideal garb we would like. I am a little concerned about the fact that when you are dealing with something that impacts people's health, you can set aside non-replicated effects that are positive in nature. I have always felt there are other ways of providing a degree of validation of positive effects that may be less than the ideal way of replicating in another laboratory. When one sees enough such positive effects occurring across a broad enough field, one begins to get the feeling that there is something here that we are going to get tripped up by if we do not understand what it is. That has been my concern in this particular field.

Dr. Joseph Morrissey, Motorola:

Thank you for your comment. I believe the approach to take here by any scientific body and which has being taken by a number of scientific bodies, the IEEE SCC28, the NRPB in the UK, and in a variety of others, is to look at the entire playing field of results. The three reviews that I showed you, the Health Canada review, the Brusick review, and the publication by ICNIRP, looked at the entire playing field and did an evaluation of the entire literature base, and the results are predominantly negative. Now again, this is not to just disregard any positive finding. But in light of the largely negative findings, they have made their evaluation to indicate that there does not appear to be any apparent effect. As a scientific body, if the results are largely negative, and a positive finding does come up, one cannot ignore it. But surely it must be replicated to validate it before this can change the group's way of thinking and before it can be generally accepted in the field.

Audience member:
The Lai study used continuous wave and pulse modulated fields. In your study, did you use the pulse modulated field, or was it only continuous wave?

Dr. Isabelle Lagroye, University of Bordeaux:
I made a mistake in the last slide. Our work included only continuous wave.

Dr. Joseph Roti Roti, Washington University Medical Center:
Isabelle's particular study was just done with continuous waves, but other studies in the lab have used pulsed waves. So far the pulsed wave studies that we have done have been negative.

Dr. Robert Johnston, Australian Communication Authority:
Could you tell us why you get such variability in your data for the 10 W/kg? You are getting almost a doubling in the frequency of the micronuclei.

Dr. Graham Hook, Integrated Laboratory Systems:
The variability in the analog and TDMA experiments seem to be related with something that was particular for these experiments because the controls were also elevated.

Dr. Johnston:
In the analog 1 and analog 2, you have got quite a doubling there.

Dr. Hook:
That is what I am saying. In the analog experiments and the TDMA experiments, there is variability at 10 W/kg. But there is also variability in the controls on those experiments compared to the concurrent controls from the second experiments. So I am not sure if it is related. The fold increase is the same, but I am not sure if there is something particular to that time.

Dr. Johnston:
Why do you get such variability from study to study?

Dr. Hook:
That is inherent in the biology. I do not see that as a lot of variability.

Dr. Johnston:
What about the two studies you did with the 5 W/kg?

Dr. Hook:
We have got two studies, and you can say that there is no dose-response because you did not get it in this one. In this case we do seem to have a dose-response. We do not have enough data to really talk about whether we have got a dose-response there or not. It is two studies, and we have nothing at 1 W/kg, so that is where we are at on

this. I would like to fill in between 1 and 5 W/kg and see what we get.

Dr. Johnston:
Compared to your previous data, looking at comet assays, breaks, and aberrations, this does not seem to be consistent.

Dr. Hook:
The mechanism is unclear at the moment because it does not seem to be directly induced.

Dr. Johnston:
But that could be something within the technique.

Dr. Hook:
In the sense that we are getting a positive effect here, yes. But it has happened over eight studies, and it is only in the treated. There are concurrent controls that are kept in the TEM cell, so it seems to be related to being exposed to RF.

Dr. Johnston:
Are you crossing your TEM cells, using one as a positive control once and the other one the next time?

Dr. Hook:
No, we did not do that. Because of the setup of the TEM cells, the second TEM cell in the series had to be the control in all cases. No, we did not reverse them.

Dr. Johnston:
It might be worth looking at.

Audience member:
I know that it was a lot of work to collect so much data from so many different experiments. The question I would like to ask is, are all these data collected from only one blood sample, one donor?

Dr. Hook:
No. Within any technology the donors the same, but there were different donors. For the analog experiments, that was the same donor. TDMA, CDMA, and PSC had the same donor for that experiment, but there were different donors. For each experiment, the blood came from one donor. For each technology, that donor was the same in both experiments. But there was a different donor used in one, in the PCS, if I remember correctly.

Mr. Kwok Chan, US Federal Communications Commission:
I have a quick question about the temperature increase because we found it difficult to have SAR levels up to 2 or 3 without a temperature increase locally at the very location where the cells are. How did you avoid this?

Dr. Hook:
We had two temperature probes. The first one was placed in the middle of the tube, which we had been told from the dosimetry was a peak area. We also had a probe that was set at the bottom, where the cells were. That probe was always 37±1°C. We did not see any temperature increase that we could measure using our probe.

Mr. Chan:
Did you use any cooling?

Dr. Hook:
I did not give the details of it, but the temperature was controlled. There is heating, but there is also cooling. We maintain it at 37°C using the fans and the water jacket.

Dr. Ross Adey, University of California at Riverside:
I would like to ask some questions and make some comments about this 10 W/kg to induce this phenomenon. As first reported in 1990 in microwave workers in Yugoslavia, they saw this micronucleus phenomenon with the estimated mean field strength of about 50 mW/cm^2. They observed this in workers exposed for 15-25 years. My questions is, how relevant is 10 W/kg for 24 hours as a measure of what may be a much more subtle cumulative process? These Yugoslav workers investigated the same phenomenon in vinyl chloride workers. In the vinyl chloride workers, there is this predisposition to cancer in general and also to leukemia. So the question of the time scale of the observation seems dubiously relevant to me in this context.

Dr. Roti Roti:
It sounds like that is a very good point to lead off the philosophical discussion of what these results are all about.

Audience member:
I would like to ask a question about the cell culture system for those of us who are not familiar with the lymphocyte cultures. At the bottom of 5 ml of media, are these cells reasonably well oxygenated and healthy? If you take them out after 24 hours, are they viable?

Dr. Hook:
Yes, they are quite viable. In fact, they have to go through a cell division before we can do this assay. That is part of the measurement, how fast they are replicating. The data we have from these studies show that the cells were not adversely affected by the cell culture situation we had in this case.

Audience member:
Are the oxygen levels comparable to other cells in a monolayer culture in a flask or dish?

Dr. Hook:
It is hard for me to say what the oxygen levels are. The medium is certainly capable

of maintaining the cells, and they grow quite well.

Dr. George Carlo, Wireless Technology Research:
This is addressed to everyone on the panel. If you look at the simplest form, we have a number of validated assays looking at genotoxic effects that are negative. We also have the micronucleus findings that are positive, at least in some experiments. If you look at that in a general context, you have a question something like this: What are the chances that the genetox battery is the correct interpretation and the micronucleus assay is just plain wrong? Or the alternative is that the micronucleus assay may be picking up something that the standard genetox battery cannot. Maybe each of the panelists might address that.

Dr. Hook:
Each of the assays is measuring slightly different endpoints, in this case, different cell lines or different exposure scenarios. So they are different in that case. Is one right and the other wrong? I do not think so. The data we have are valid in both sets of tests. The genotoxicity tests were repeated twice for each laboratory, as I said, the first round using an analog system that was actually done by another laboratory, so there were four independent assays there. Those are clearly negative studies. We have repeated the micronucleus assay four different times. It is positive. Which one is right or wrong? You have to tell me what criteria you want me to base right or wrong on. Those are both valid sets of data. What it tells us is we do not know what the mechanism is for the induction of the micronucleus. It is not a simple one. It is not a straightforward mechanism.

Dr. Donald McRee, Wireless Technology Research:
My interest, rather than the genetox battery compared to the micronuclei assay, is the SCG in comparison to the micronuclei assay. I have always thought of micronuclei being formed by bits of DNA, and the SCG assay is supposed to measure DNA breaks or DNA damage. I have been educated a little bit that micronuclei formation can be in the spindle replication stage. We hope to be able to generate some data. We actually made some slides that we hope to be able to analyze that can tell us whether it is pieces of DNA or the function of the spindle in replication that is causing the effect. Graham and Ray Tice both know more about this than I do. But to me that is the more interesting comparison in terms of what is going on in the mechanism, that sort of thing. It could be purely a difference in thermal gradience, but it would be really interesting because of all the experiments that we have run at these higher levels, we have gotten a very repeatable response in terms of frequency of the micronuclei.

Dr. Morrissey:
As Dr McRee's last slide listed, there are certain questions that need to be addressed before one can really move on and ask whether one is right or wrong. I think those questions were right on target, looking at potential thermal problems and doing a replication as Joe has suggested. Those things need to be addressed first.

Dr. Roti Roti:
I would like to return to the beginning of this Colloquium when you mentioned the bottom line is health hazards. I would like to introduce at this point in our discussion that to really assess health hazards, on the table should be a clear cut mechanism of interaction. I think we can all agree we do not have a clear cut mechanism of interaction, possibly because electromagnetic fields at these levels have something that we call a weak biological interaction as opposed to a strong and robust biological interaction. What I see these data as are perhaps an opportunity to delineate what might be a mechanism of interaction. I think it may be more important to establish that point, and after that is established, we can ask if there is a health hazard associated with whatever that interaction is.

Dr. Carlo:
One follow-up question. Do any of you have concern with regard to these 200 million people who are using the technology while we in the scientific community are sorting this out?

Dr. Morrissey:
There are quite a few studies that have been either recently presented or are ongoing, and still the majority of these studies do not cause groups like NRPB or the IEEE SCC28 to be concerned.

Dr. Carlo:
To be fair, none of these groups have seen any of these data. I think it is not fair to be pointing to those deliberative proxies that did not look at any of the new data we have seen here today.

Dr. Morrissey:
We saw reports of both effects and no effects. Again, these need to be replicated and addressed.

Dr. Roti Roti:
I am glad you asked that question in a way. You were nice enough to invite me to speak at the first symposium, and we had just barely begun our work and had completed no studies at that point in time. Since then we have completed many studies, published 12 of them, and we are working as fast and as hard as we can. I really think that the cart and the horse should be in the right order.

Dr. McRee:
I think it is interesting that we did not see a micronuceli increase at 1 W/kg average SAR which had some cells that were being exposed as high as approximately 2 W/kg. If you want to apply that to cell phone use, most of the information you see that indicate that brain cells are not getting but around 1 W/kg. I think the data do not indicate that at that level we really see an increase in micronuclei. At this stage going from an in vitro system to an in vivo system is a stretch anyway, but I do not see how even if you try to directly apply it that it seems to be a problem on the basis of the data

we have now.

Dr. Roti Roti:

A scientific answer to your question is that many of the genetoxic assays that turned up negative were based on population means. With the comet assay you tend to look at the means, you do not necessarily look at the tail 1-2% of the distribution. Whereas with micronucleus and mutation assays, you are looking at the behavior of the small percent of the population that does change. You are really looking at different endpoints, and you are asking a different proportion of the cells to do something that you can measure.

Dr. Hook:

Just in terms of the SCG assay, we do in fact look at the distribution to see if there are any kind of outlying populations. When we have a negative in the SCG, it is negative in that sense too. Just to preface, I am not thinking that what we have does or does not have relevance. But these are hazard identification assays, and to extrapolate them to mean that there is no effect in humans because we did not see something at 1 W/kg is as misleading as saying that because we got something at 10 W/kg that there is a problem. It is giving you some data by which you can then go ahead and do further studies to evaluate what this means in the human exposure scenario. This is unstimulated blood in a test tube in a TEM cell. This is not a person using a cell phone. It is too many leaps to extrapolate from these data. It tells you there is something we need to look at. One of the best ways of doing that is to understand the mechanism of action. That is where you can then start extrapolating to humans, saying is that a plausible mechanism in the human exposure scenario.

Audience member:

For those of us that are not that familiar with the micronucleus assay, is there anyone on the panel who could elaborate on how this assay has served risk assessment in the past? In other words, the idea is now you would go on and maybe do some additional experiments. Is there a paradigm in which the micronucleus assay has worked well to identify hazard and then to go on to develop some ideas about human health risk assessment?

Dr. Hook:

That is a pretty general question. Most things are tested in a battery or a series of assays. The particular assay we used here is a little unique. The more general assay that is used is the chromosome aberration assay rather than the micronucleus. There is a push to using the micronucleus assay. The simple answer here is the next step is to look for the same endpoint in in vivo systems as we had seen in in vitro. If you are looking at the health paradigm, then the next move is to look at the in vivo exposure scenario and try to measure the same endpoint.

Dr. Lundquist:

One gentleman made the comment that he did not want to talk about some of these things except in context of specific mechanisms, so I thought I would say a few words about mechanism which I did not mention in my presentation this morning quite deliberately. Perhaps I should say that I get contacted by people from all over the world who have felt that they are being adversely affected by electromagnetic fields of one kind or another, very often microwaves, though not exclusively. So I have the experience of a clinician to a degree in a terms of finding out what their experiences have been in whatever detail they want to tell me. I find that I am often referring them to physicians in the field of environmental medicine who deal with two types of problems, electrical sensitivity and multiple chemical sensitivity. These do seem to be related. My opinion based on what I have learned is that the exposure to the radiation ends up altering one's sensitivity to chemicals. This suggests that one of the things that the radiation does is to make some modification either in chemical reaction rates which is certainly possible. Or based on what I said this morning about the Maxwell stress tensor, you end up having forces applied to biological molecules that alter their configuration which certainly could have biological effects. We know that proteins, for example, can assume any of several different configurations. There is good reason to believe that in one configuration they will undergo certain types of reactions, and in another configuration they will undergo certain other types of reactions. So you might very well have a biological or health effect as a consequence of a change in configuration that such a molecule might undergo.

Dr. Roti Roti:

Thank you very much. This gives me a chance to tell you some of the other work we have done that we do not have time to present at this Colloquium. Obviously the nucleus consists of DNA, but the DNA is not freestanding within the nucleus. It exists in the context of nuclear proteins which regulate accessibility to that DNA for various functions, including gene expression, DNA repair, and so forth. This paper has been submitted. We have found no effect that could be measured—again all those possible effects, plausible though they might be, if they are going to be relevant, they need to be measurable at some level. We began to attempt to do that for nuclear protein changes and found nothing. We have also tried to look at chemical reaction rates, and we tried to focus on the one that would be most important for cancer risks, that is, those reactions that relate to oxidative stress. Those studies are ongoing and we hope to have an answer fairly soon as to whether or not any measurable effect on free radical processing can be found. In some data I will present tomorrow, we have made direct measurements of transformation frequencies following exposures to these fields. We have conducted a 2-year bioassay the results of which are in the process of coming out. Therefore I do not think we are remiss and we are not brushing things off lightly. We are considering all possible things and trying to measure them within the level of our resources.

Dr. Michael Repacholi, World Health Organization:

I would like to reiterate what Graham said, that something found in vitro does not necessarily mean it will be found in vivo. Certainly WHO would require that there be either a mechanism identified or that the effect be found in vivo so that you have more comfort with this being extrapolated into humans. I would like to come back to Ross Adey's question, because it is an interesting one. Is there the possibility that you could have an accumulation of micronucelus formation over a long period which would then be found in workers after a number of years? I am not terribly familiar with this, but I think that micronuclei do stay in the cell for numbers of generations. Could they accumulate? If you did have micronuclei formed by lower level fields, could they accumulate and be found a number of years later?

Dr. Hook:
There is a turnover rate for all the cells we are talking about, lymphocytes and red blood cells, so they would accumulate within that life span, and then they would disappear as the cells disappeared. In general you would think that in the human population they would be at some saturated level and that is what you would see, if they were being continuously exposed. If the exposure stopped, then that level would go down, you would think. In lymphoyctes there are some segregated populations, and some of them seem to have life spans of decades. Some have a shorter life span. In peripheral blood, which would be another endpoint you could use, you could not really use it in humans because the red blood cells with micronuclei disappear in humans.

Dr. Repacholi:
What lymphocytes stay around for decades?

Dr. Hook:
I do not know if it is a type of lymphocyte in any particular case, but in my understanding there are some lymphocytes that last a long time.

Dr. Repacholi:
The other question is, Dr. Roti Roti, are you going to do your experiments in lymphocytes to complement Graham's results?

Dr. Roti Roti:
That is what I would like to do. I think that is something that is under discussion. That would be what I would recommend we do, is that we actually run these experiments with each other's systems. We would obviously have to trypsinize those cells and expose them in a pellet, but that is feasible. We would have to make some special holders which we have actually done. These data are actually in Dr. Goswami's paper, but we have carried out transcription factor activations which are actually done in vitro. So we made some exposure systems where we could actually expose these solutions or cell preps in microfuge tubes, so that kind of exposure system is available. We would have to do something like that with the lymphocytes. We have a few engineering hurdles to overcome before we can actually run off and do these experiments. That is one of the reasons I did the pilot is to justify the engineering that we will probably have to go through to do this correctly. What you have suggested is exactly what I would like to see happen and would recommend personally.

APPENDIX C-4

IN VIVO STUDIES

Audience member:
What is the time table for the Australian-funded replication study?

Dr. Michael Repacholi, World Health Organization:
It is in progress now, will be completed within the next 2 years, and then will be published a year after that.

Dr. Vijayalaxmi, The University of Texas Health Science Center:
Dr. Hook, in the data table between ILS and the University of Washington, there is not much difference between percent DNA and tail moment, but the University of Washington study scored twice the tail length as your study. Is that because you used the meta-analysis system and the University of Washington used the micrometer, or did you use the same system?

Dr. Graham Hook, Integrated Laboratory Systems:
The majority of that difference is due to the fact that one is measured in pixels and the other is measured in microns. Guesses are being made when they are converting from pixels to microns. But if you can make estimates, through guessing about what a pixel size should be and that conversion is made, they come out to be the same. Overall, there is some increase in the tail length in the University of Washington study but it is small.

Dr. Vijayalaxmi:
Did the University of Washington use the meta-analysis system?

Dr. Hook:
Yes. The University of Washington used the same identical image analysis system as what I used, including the same camera. The program allows you to put in a value

which converts pixel length into microns. If you do not convert pixel length into microns, the data will be evaluated as pixels and if you do then the data will be evaluated as microns.

Dr. Repacholi:
Dr. Hook, can you indicate what Henry Lai says about the differences between the results and if Dr. Lai agrees with your conclusion that there is really no difference in results of this comparison?

Dr. Hook:
Dr. Lai has made no comment on this comparison, probably because has not seen it.

Dr. Marjorie Lundquist, Bioelectromagnetic Hygiene Institute:
I have a comment about the statistical evaluation where the significance level chosen was the usual 0.05. I feel that corrections for multiple comparisons are not being handled correctly in the statistical community at the present time. Originally, statistics was an attempt to objectify decision making, especially in the biological sciences where there is a high degree of variability. If you did not have something like statistical techniques to apply, you are basically making objective evaluations which can very greatly from one investigator to another. The attempt that was made originally was to choose your significance level at the outset before you had run your experiment which would be the subjective decision made and you made it before you had any data available. Once this decision was made, you had to stick to the results of the statistical analysis. If they were statistically significant you had a real difference. The point is, today scientists might decide that they have a reason to reject the statically significant results because of multiple comparisons, which takes us back to making a hidden subjective evaluation that is clothed in statistical language and looks objective but is really fairly subjective.

Dr. Hook:
I would like to make a general comment to that statment. The corrections were part of the statistical package put together and included in the protocol prior to the initiation of the study and an important point is that the statistical procedures be defined in a protocol before the study is started. As with all other aspects of the protocol and with the corrections in this case, they were put into the protocol by the researchers themselves and considered to be an important part of their statistical analysis. The statistical conclusions and the final conclusions on any of these data and any toxicological data used in risk assessments are that there is an important aspect of review to be performed by experienced scientific personnel. Nothing should ever be based purely on a statistical call, negative or positive.

Dr. Repacholi:
I would like to comment in addition to Dr. Hook's comment. Various comparisons are done to look at internal consistency within the study to make sure that any chance events are not going to dominate in the result. What is being looked for is internal consistency within a study to make sure that it is a true negative or true positive result.

Dr. Repacholi:
I just want to say how good it is to discuss something from results we have presented by the authors instead of reading about studies for the first time in the press. I cannot tell you how much damage it does to the field when authors go straight to the press rather than through peer-reviewed publications and then allowing scientist to provide peer review to these studies so that everyone can be at the same level. I would encourage scientists to reconsider, more than once, going to the press with publications instead of allowing the scientific community to critique them first.

Dr. Vijayalaxmi:
There is a difference between the studies that have been discussed here this morning and the study I did. The main difference is the discussion that centers around brain cells in rats in the morning, where as in my study I used human blood lymphocytes. I used 2450 MHz pulsed wave RF in vitro for 90 minutes. The comet slides were made immediately after exposure, and then the same samples kept at 37°C for 4 hours. The difference I want to emphasize is that Singh has used brain cells and I have used human blood lymphocytes for my experiment.

Dr. Repacholi:
We have someone who is replicating the *pim*-1 study, Tim Kusch, and I would like to invite him up to speak briefly about it.

Dr. Tim Kusch, World Health Organization:
The follow-up to the Repacholi study is called a follow-up rather than a replication because, as Mike pointed out in his presentation, there were deficiencies, particularly in relation to dosimetry, in the first study. There was also only an attempt to expose the animal to one dose of SAR. In our current project we have five doses from 0.25 to 4 W/kg. All of the problems associated with the first study, such as disease and obesity, have been taken into account within the statistics this time. We started the exposures a month ago. The gestation of the experiment is about 12 months. It is a 2-year study and perhaps in about 2.5 years' time our manuscript should be ready.

Dr. Repacholi:
When do you expect the last group of mice to go into the exposure?

Dr. Kusch:
The last group of mice were introduced 3 weeks ago.

Dr. Repacholi:
What about p53 study?

Dr. Kusch:
The National Health and Medical Research Council did not in fact fund the p53 study, but the door is open for that to occur. If we were to pursue that, that would have to start at the end of the first study anyway.

Dr. Donald McRee, Wireless Technology Research:
What if you do find a replication or a close replication or increase in lymphoma? What does this mean in terms of extrapolating from a transgenic animal to a human? Have you given much thought to that?

Dr. Repacholi:
I asked IARC about that, and they said they would accept transgenic mouse data in the assessment of carciogenicity of a physical or chemical agent in the same way that transgenic mice are now used for screening with new drugs and chemicals.

Dr. Joe Eavis, University of Bristol:
You described an experiment in which you treated animals with DMBA and then exposed them to radiofrequency radiation at five different SARs. You said you saw increased tumors above 1.4 W/kg meaning you had increased tumors at the two higher SARs, 2.2 and 3.5 W/kg. Will you give us more details on that?

Dr. Rachid Anane, University of Bordeaux:
I will be presenting these results in a more detailed fashion at BEMS on Thursday. We did two experiments in two different seasons, one from May to July and the second from September to December. In the first experiment we exposed animals to 1.4, 2.2, and 3.5 W/kg. For the second experiment we chose one common level of exposure. We exposed five groups: one control cage with an incubator, a sham exposure, and three exposed groups. One group was exposed to 1.4 W/kg, and the two other groups were exposed to 0.1 and 0.7 W/kg. I have shown in my slides that for the 1.4, 2.2, and 3.5 W/kg exposures, we had an increase in the number of malignant tumors for the groups. For the second we had a decrease of tumors for the groups exposed to 0.4 W/kg. There was no difference between the sham and the two groups exposed to 0.1 and 0.7 W/kg.

APPENDIX C-5

EPIDEMIOLOGY

Dr. Marjorie Lundquist, Biolelectromagnetic Hygiene Institute:
I have a question regarding our comparison between Norway and Sweden with respect to the GSM and NMT telephones to see what might be similar in the two countries. I noted that tingling/tightness seemed to be a distinct finding suggesting a stronger effect in the GSM phone than in the NMT phone. Can you explain what exactly that means?

Dr. Gunnhild Oftedal, Norwegian Institute of Science and Technology:
Tingling/tightness is a sensation in the facial skin. You are right; there was a similar effect in both countries but not significant enough in the same direction.

Audience member:
Dr. de Seze, could you provide a sampling frame for the melatonin study and the number of people involved?

Dr. Rene de Seze, CHU Nimes:
Yes. The number of people were 18 and 19 in each of the experiments. The sampling frame consisted of one sampling each hour during the night and one sampling every 3 hours during the day. Volunteers were asleep and blood was drawn through a catheter set in their arms.

Dr. Lundquist:
I would like to make a comment about your last visual. I had an extensive discussion with Dr. Alan Preece in the UK after his paper was published indicating no finding of short-term memory loss. I would like to recommended that if you were going to be doing experiments of this kind, it would be wise to allow about 15 minutes for the effect to establish itself before you test for its presence. I think the reason Dr. Preece got negative effects was that he looked for a biological effect before it appeared.

Dr. de Seze:
My laboratory plans to look at chronic exposures on the same scale that I presented in this speech. That means an exposure time of 2 hours each day for 5 days per week for 4 weeks.

Dr. Joseph Morrissey, Motorola:
I would like to ask Mr. Eavis about the change in choice reaction time. From reading the paper, it was not the greatest difference that was reported, and it seemed like a very tight set of data. I would like to know if you have the raw data for this measurement or the standard errors.

Mr. Joseph Eavis, University of Bristol:
I do not, but Dr. Preece certainly would have. Dr. Preece will be giving an extended version of this presentation on Thursday at the BEMS meeting, including what was covered today in addition to the cerebral blood.

Dr. Morrissey:
Are the raw data listed in the paper?

Mr. Eavis:
They are not listed in the paper, but Dr. Preece does have the raw data.

Maria Feychting, Karolinsky Institute:
I would like to comment on the number of cases in the Swedish study on mobile phone use and cancer. I understand that Lennart Hardell is not here to answer this and that Kjell Mild has only been involved in the exposure assessment. I still feel that it is necessary to point out the area where Hardell has performed this study (the Stockholm area during 1995-1996 and the Uppsala area during 1994-1996) in those aged 20 to 80. If you look in the Swedish cancer registry you would have identified 862 cases of brain tumors during this period, 431 cases of gliomas, and 264 cases of mesotheliomas. In the Hardell's study, he identified 270 brain tumor cases, which is only 31% of all the available cases. One possible explanation could be that it is a rapidly fatal disease he is studying. But when we look at the survival time we find that 77% of the cases are still alive at 6 months after diagnosis and 65% of the cases are alive 1 year after diagnosis. Dr. Hardell has not included all cases as he has written in the publication. We need to know the selection procedure to be able to evaluate possible selection bias in this study because 31% of the cases is quite a small proportion of occurred cases. This could have an impact of the validity of the study.

Audience member:
Mr. Muscat, can you give a conclusion to your study?

Mr. Joshua Muscat, American Health Foundation:
Within the study population the data suggest that there is no increased risk of malignant brain tumors with cell phone use given the caveat that the number of years

these devices have been used is relatively low. My conclusion is that the case-control approach needs to be continued, looking at the association with long-term use.

Dr. Michael Repacholi, World Health Organization:
I want to emphasize that the IARC study is going to be a very large study involving at least 14 countries and a predicted number of over 7,000 cases and 7,000 controls. Therefore, if everything goes well, there will be 14 studies that can be published independently and then, due to the common protocol, results can be combined in a valid way to become a much more powerful study. Overall, this study is going to be a landmark study for this area.

Mr. Muscat:
I would like to say that it might be worthwhile to do some type of passive surveillance. If the rates of brain cancer were looked at over decades, they tend to be stable. The reason for that is there are really are not any known risk factors for brain cancer, so even though environmental exposures, smoking habits, and other individual behaviors may change, you tend to see that the rates of brain cancer remain relatively low and actually do not change much. The slight increase in brain cancer over the last 10 years is due to an increase in MRIs which is really more of a detection problem. Another approach is to look at the rates of tumors in other countries that have cancer registries because the usage of cell phones is increasing so dramatically. Also, another way to monitor health effects is to correlate the increase use of cell phone usage with brain tumor rates.

Audience member:
I would like to say that before those recommendations are followed, I believe it would behoove us to look at what we expect to see, which is pretty easy to calculate. The total rate of brain cancer would be a function of the relative risk conveyed by use in the percent of people using cell phones. When you do those calculations it is surprising how little of an effect you would see. For example, if 20% of the population were using cell phones and it conveyed a five-fold increase risk for everyone who used one, then the rate would go up by a factor of 1.4, which is not so dramatic. So it is important to know what you would expect before you do that.

Mr. Muscat:
That is an excellent suggestion. These rates are often based on death certificates which are often misclassified with respect to brain cancer. Sometimes metastatic tumors are considered primary brain tumors.

Dr. George Carlo, Wireless Technology Research:
Is there a need for a pathology review of the neuroepithelial tumors? If so, is there anything planned along those lines?

Mr. Muscat:
At this time there is nothing planned, but a larger than expected number of neuroepithelial tumors was reported in the study. The expect numbered of tumors was

unknown due to the uncommonness of these tumors and the lack of literature that describes these tumors. Of course, there is a publication bias with the respect to how frequently these types of tumors are occurring. I might mention that two of the patients had their pathologies re-read by second neuro-pathologist that came up with a different histologic diagnosis, which is not uncommon. Actually it is quite common for someone with a rare type of tumor to get a second reading of their pathology. Pathological diagnosis is not black and white, and there are difficulties in defining tumors. Overall, the finding of this one particular type of tumor is unusual, and there would be a need to do some type of pathological classification. It may not be practical to do that within our own study setting, but I recommend in future studies that histology should be looked at by at least one or two different readers.

Dr. Carlo:
Dr. Mild, do you and Dr. Hardell have a time table for the analysis of the 1,000 or so cases that have already been gathered?

Dr. Kjell Hansson Mild, National Institute for Working Life:
What needs to be done is to collect more data to reach 1,200 which could take another 6 months or so. Dr. Hardell and I hope to have something for the BEMS meeting in Munich next year, but it might be tight.

Dr. Repacholi:
Dr. Mild, did the study not report some positive studies and positive findings?

Dr. Mild:
That is not how we presented it. On the whole it is negative. There is a possibility that there might be something with the sidedness, but in our study it was a non-significant increase. Overall, there is no increase between the users and non-users.

Ms. Jo-Anne Basile, Cellular Telecommunications Industry Association:
Dr. Carlo, I would be remiss if I did not take the opportunity to extend a thank you to you from the industry. You are right in saying that you have caused some sleepless nights for us, in particular over the last couple of weeks as we have heard media reports and allegations about the research. I want to thank you for hosting a truly quality 2 days of scientific discussion and debate about the issue. It gave us in the industry an opportunity to hear firsthand from the principal investigators of the research and to hear some of the give and take. For that we really do thank you; this was truly a quality presentation over the last 2 days. The other thing I want to say to you is that this industry, as you said, has not walked away from things and we do not intend to walk away from things in the future. This industry, both in the United States and world wide, is committed to ongoing research, and we will do that following Wireless Technology Research here in the United States and around the world.

APPENDIX D

HISTORY AND FACTS ABOUT WIRELESS TECHNOLOGY RESEARCH, LLC

WTR MISSION STATEMENT

The mission of Wireless Technology Research, LLC (WTR) is to identify any public health problems that follow from the use of wireless technology and to make timely recommendations to correct any public health problems that are found.

Support of the independent WTR by the wireless industry underscores the industry's ongoing commitment to assuring that wireless technology is used safely.

The methods developed and followed by WTR are consistent with standard product safety evaluation and post-market surveillance requirements of the US Food and Drug Administration (FDA).

The voluntary and significant support of this program by the wireless industry has obviated the need for regulatory intervention by US government agencies.

KEY DEVELOPMENTS IN THE HISTORY OF WTR

Following allegations of increased brain cancer risk from the use of cellular phones on the Larry King show in early 1993, the wireless industry committed itself to a research program to look into the allegations. Parameters of that commitment were made public in February 1993 and included a timeframe of 3-5 years and funding support between $15 and $25 million.

In April 1993, the Scientific Advisory Group (SAG) was established to formulate a program that was responsive to the public health risk allegations and to advise the wireless industry of such.

In May 1993, the SAG program of surveillance and intervention was presented to the FDA and approved in concept, with the charge of implementation given to the

SAG.

In June 1994, the SAG became aware of and took action on two important pieces of scientific information: The first was evidence from Europe that cellular phones could interfere with implanted cardiac pacemakers; the second was a report by Drs. Henry Lai and N. P. Singh of the University of Washington that low-power radiofrequency (RF) radiation could cause DNA damage.

In August 1994, the SAG published its peer-reviewed Research Agenda aimed at filling in scientific data gaps that needed to be filled before informed judgments about the public health impact of wireless phones could be definitively assessed.

In December 1994, the US General Accounting Office issued a report praising the work of the SAG but suggested that greater credibility for the activity would be gained if the work were funded through a blind trust. In response to this recommendation, the wireless industry and the SAG agreed to form the independent entity, Wireless Technology Research, LLC, with funding for the group directed by the industry through a one-way deposit-only escrow fund. An Audit Committee was put in place to oversee an annual independent audit of the financial practices of WTR and to report their findings annually to the industry.

In November 1995, WTR, along with the International Coordinating Committee on Wireless Health Research, sponsored the first comprehensive colloquium on the state of science regarding public health risks from wireless technology in Rome, Italy with participation of more than 100 top scientists from around the world.

In September 1996, WTR issued its comprehensive recommendations for solving the problem of digital phone interference with implanted cardiac pacemakers

In December 1998, WTR-sponsored scientists began releasing results of toxicology and epidemiology studies conducted under the WTR Research Agenda.

HOW WTR DECISIONS ARE MADE

WTR scientific staff are ultimately responsible for making evaluative decisions within the program. However, it is the policy of WTR to receive input and peer review from a wide range of scientific and government groups.

The WTR decision-making framework follows the four operating questions of the public health paradigm, and all information potentially useful in identifying public health problems is evaluated against that template.

When a decision about research or intervention is made, WTR initiates appropriate briefings with government and affected industry groups.

PEER REVIEW BOARD

At the beginning of the program in 1993, a blue-ribbon Peer Review Board (PRB) was assembled to perform two specific functions:

1. To conduct scientific peer review of program work products including the Research Agenda, study protocols, and research reports, and
2. To respond to specific questions posed by program personnel from time to time.

In early 1994, the Harvard Center for Risk Analysis was charged with the task of coordinating the peer review activity.

The process now involves transmittal of protocols, research reports, or lists of specific questions from WTR to the Harvard Center for Risk Analysis, which in turn coordinates peer review activity among members of the PRB independent of WTR. The Harvard Center for Risk Analysis consolidates the PRB comments and transmits them in writing to WTR.

The Peer Review Board was comprised of the following individuals:

Larry E. Anderson, Ph.D.
Bioelectromagnetics Program Manager
Departments of Biology and Chemistry
Battelle Pacific Northwest Laboratories

Patricia Buffler, Ph.D., M.P.H.
Dean, School of Public Health
University of California at Berkeley

Sir Richard Doll, F.R.S., F.R.C.P.
Emeritus Professor of Medicine
Oxford University, England

Carl H. Durney, Ph.D.
Professor, Electrical Engineering Department
University of Utah

Joe Elder, Ph.D.
Special Assistant
National Health and Environmental Effects Research Laboratory
U.S. Environmental Protection Agency

John D. Graham, Ph.D.
Director, Center for Risk Analysis
School of Public Health
Harvard University

Saxon Graham, Ph.D.
Professor Emeritus
State University of New York at Buffalo

Don R. Justesen, Ph.D.
Professor of Neuropsychology, School of Medicine
University of Kansas
Research Service, USVA Medical Center

Richard R. Monson, M.D., Sc.D.
Professor of Epidemiology, School of Public Health
Director, Educational Resource Center for Occupational Safety and Health
Harvard University

Susan W. Putnam, Sc.D.
Research Associate in Environmental Policy
Center for Risk Analysis, School of Public Health
Harvard University

Asher R. Sheppard, Ph.D.
Consultant
Asher Sheppard Consulting

Dimitrios Trichopoulos, M.D., S.M.
Director, Harvard Center for Cancer Prevention
School of Public Health
Harvard University

Peter A. Valberg, Ph.D.
Cambridge Environmental, Inc.

Gary M. Williams, M.D.
Chief, Division of Pathology and Toxicology
Director of Medical Sciences
American Health Foundation

Sheldon Wolff, Ph.D.
Vice Chairman and Chief of Research
Radiation Effects Research Foundation

WTR has provided, over the course of the program, periodic progress reports to the PRB as a group. Reports or minutes from these group meetings are also transmitted to WTR by the Harvard Center for Risk Analysis in writing.

US GOVERNMENT INTER-AGENCY WORKING GROUP ON RF

Because the activities of WTR are of interest to a wide range of government agencies, WTR has regularly met with the formal Inter-Agency Working Group (IAWG) on RF. The IAWG, which is comprised of top officials and scientists from agencies such as the FDA, the Federal Communications Commission, the Environmental Protection Agency, and the Occupational Safety & Health Administration, is coordinated through the FDA.

WTR and the IAWG meet periodically to share information and for the IAWG to provide peer review input to WTR.

The IAWG is comprised of the following individuals:

Elizabeth D. Jacobson, Ph.D.
Deputy Director of Science
Center for Devices and Radiological Health
U.S. Food and Drug Administration

Russell D. Owen, Ph.D.
Chief
Radiation Biology Branch
Center for Devices and Radiological Health
U.S. Food and Drug Administration

Larry W. Cress, Ph.D.
Medical Officer
Radiation Biology Branch
Center for Devices and Radiological Health
U.S. Food and Drug Administration

Donald E. Marlowe
Director, Office of Science and Technology
Center for Devices and Radiological Health
U.S. Food and Drug Administration

Howard I. Bassen
Chief, Electrophysics Branch
Center for Devices and Radiological Health
U.S. Food and Drug Administration

Joanne Barron
Regulatory Officer, Office of Compliance
U.S. Food and Drug Administration

Larry G. Kessler, Sc.D.
Director, Office of Surveillance and Biometrics
Center for Devices and Radiological Health
U.S. Food and Drug Administration

Ronald G. Kaczmarek, M.D., M.P.H.
Medical Officer, Office of Surveillance and Biometrics
Center for Devices and Radiological Health
U.S. Food and Drug Administration

Joe Elder, Ph.D.
Special Assistant
National Health and Environmental Effects Research Laboratory
U.S. Environmental Protection Agency

Norbert N. Hankin
Environmental Scientist
Office of Radiation and Indoor Air
U.S. Environmental Protection Agency

Gregory W. Lotz, Ph.D.
Chief, Physical Agents Effects Branch
Division of Biomedical and Behavorial Science
National Institute of Occupational Safety and Health

Robert F. Cleveland, Ph.D.
Physical Scientist
Office of Engineering and Technology
Federal Communications Commission

Robert Curtis, Ph.D.
Director
Division of Program Support
Occupational Safety and Health Administration

Janet Healer, Ph.D.
Bioscientist
National Telecommunications Information Administration
Department of Commerce

Mary Wolfe, Ph.D.
Associate Coordinator
EMF Hazard
National Institute of Environmental Health Sciences

Martha Linet, M.D.
Medical Officer
Radiation Epidemiology Branch
National Cancer Institute
National Institutes of Health

INTERACTION BETWEEN WTR AND FDA

WTR and FDA scientists and officials communicate regularly and share information about both science and policy on an ongoing basis.

WTR invited FDA scientists to assist in the analysis of the data gathered in WTR's clinical study of interference between wireless phones and implanted cardiac pacemakers, and scientists from both organizations have collaborated in scientific publications.

WTR has briefed FDA scientists and officials on the findings of toxicology and epidemiology studies currently under peer review.

WTR INTERACTIONS WITH THE WIRELESS INDUSTRY

WTR interfaces with members of the wireless industry through four well-defined mechanisms that do not compromise the independence or the scientific integrity of

WTR activity.

The Joint Review Committee/CTIA - When questions about the safety of wireless instruments were first raised in 1993, a committee of executives from both the manufacturer segment of the industry and the carrier segment of the industry was formed to oversee the research program to which the industry had publicly committed. CTIA has evolved as the follow-on organization to carry on the role of that committee. WTR held regular briefings with the Joint Review Committee and subsequently with CTIA to keep them informed of the progress and scientific findings of the program. When positive findings are identified, such as the case with digital phones interfering with implanted cardiac pacemakers, WTR advises the industry early in the process so that the industry can work toward solutions in conjunction with WTR and government.

The Audit Committee - With the formation of the independent WTR in 1995, an audit committee was formed to oversee an annual audit of WTR financial practices and to make an annual report to the industry of its findings. WTR interacts with the independent auditors annually and is responsive to requests for information from the audit committee.

One-way Deposit-only Escrow Fund - The primary industry funding of the WTR program is through an escrow mechanism that shields WTR from industry influence on its science. The industry makes regular payments to the fund and WTR distributes the fund according to its independent internal budget structure.

Peer Review Requests - WTR seeks peer review scientific input on a periodic basis from scientists within the wireless industry. This input is especially important where proprietary information is of interest and where only the industry has access to it. In addition, the Risk Management Research function of WTR necessarily required input from industry scientists where risk management options such as technology modifications were of interest.

WTR PARTICIPATING SCIENTISTS

Consistent with the WTR program criteria regarding its support of only the highest quality scientific work and independence from industry influence, WTR makes every effort to contract with scientists who have proven scientific accomplishments and have impeccable reputations regarding scientific integrity.

All studies conducted pursuant to the WTR Research Agenda are required by contract to follow Good Laboratory Practices, Good Clinical Practices, and Good Epidemiology Practices, which encompass the most rigorous quality control procedures. In addition, protocols are peer reviewed prior to being implemented in the field or laboratory, and the research reports which follow from those protocols are also peer reviewed through the Harvard Center for Risk Analysis Peer Review Board. Finally, WTR participating scientists are required by contract to submit their work for publication in the peer-reviewed scientific literature.

WTR dosimetry scientists include the following individuals:

Chung-Kwang Chou, Ph.D.
City of Hope National Medical Center

Jeffrey Wong, M.D.
Division Chair, Radiation Oncology
City of Hope National Medical Center

John McDougall, B.S., B.A.
Department of Radiation Research
City of Hope National Medical Center

Kwok Chan, M.S.
Department of Radiation Research
City of Hope National Medical Center

WTR toxicology scientists include the following individuals:

In Vitro Toxicology

Bacterial Cultures (*Salmonella/E. coli*)

E. Maria Donner, Ph.D.
Integrated Laboratory Systems

Edward S. Riccio, B.S.
SRI International

Richard A. Winegar, Ph.D.
SRI International

Mammalian Cell Cultures (Mouse Lymphoma Assay)

Raymond Tice, Ph.D.
Vice President, Research and Development
Integrated Laboratory Systems

Pamela Lee, B.A.
SRI International

Richard A. Winegar, Ph.D.
SRI International

Human Blood Lymphocytes (Chromosomal Aberration Assay)

E. Maria Donner, Ph.D.
Integrated Laboratory Systems

James Bakke, B.S.
SRI International

Richard A. Winegar, Ph.D.
SRI International

Human Blood Leukocytes (Single Cell Gel Assay/Comet Assay)

Raymond Tice, Ph.D.
Vice President, Research and Development
Integrated Laboratory Systems

Human Blood Lymphocytes (Micronucleus Assay)

Raymond Tice, Ph.D.
Vice President, Research and Development
Integrated Laboratory Systems

In Vivo Toxicology

Rat Brain Cells (Single Cell Gel Assay/Comet Assay)

Raymond Tice, Ph.D.
Vice President, Research and Development
Integrated Laboratory Systems

Henry Lai, Ph.D.
Bioelectromagnetic Research Laboratory
University of Washington

Narendra Singh, Ph.D.
Bioelectromagnetic Research Laboratory
University of Washington

WTR epidemiology scientists include the following individuals:

Case-Control: Acoustic Neuroma

Joshua Muscat, M.P.H.
American Health Foundation

Michael Huncharek, M.D.
Massachusetts General Hospital

Mark Malkin, M.D.
Memorial Sloan-Kettering Cancer Center

Alfred Neugut, M.D., Ph.D.
Associate Professor of Clinical Medicine and Public Health (Epidemiology)
Columbia University School of Public Health

Lloyd Alderson, M.D.
Rhode Island Hospital

Fred Golumb, M.D.
New York Medical Center

Case-Control: Brain Tumors

Joshua Muscat, M.P.H.
American Health Foundation

Michael Huncharek, M.D.
Massachusetts General Hospital

Mark Malkin, M.D.
Memorial Sloan-Kettering Cancer Center

Alfred Neugut, M.D., Ph.D.
Associate Professor of Clinical Medicine and Public Health (Epidemiology)
Columbia University School of Public Health

Lloyd Alderson, M.D.
Rhode Island Hospital

Fred Golumb, M.D.
New York Medical Center

Cohort: Cause-Specific Mortality

Kenneth Rothman, Dr.P.H.
Senior Epidemiologist
Epidemiology Resources, Inc.

Nancy A. Dreyer, Ph.D.
President
Epidemiology Resources, Inc.

Jeanne E. Loughlin, M.S.
Epidemiology Resources, Inc.

WTR clinical scientists include the following individuals:

Multi-center Pacemaker Study

David L. Hayes, M.D.
Director of Pacing Laboratory
Department of Cardiology
Mayo Clinic

Paul J. Wang, M.D.
New England Medical Center

N.A. Mark Estes III, M.D.
New England Medical Center

Dwight W. Reynolds, M.D.
Vice Chief, Cardiovascular Section
University of Oklahoma Health Sciences Center

Risk Management Study on Pacemaker Interference

Roger G. Carrillo, M.D.
Cardiovascular Surgeon
Department of Thoracic and Cardiovascular Surgery
Mount Sinai Medical Center

AD HOC COMMITTEES

WTR convenes specialized scientific panels and committees to assist with specific questions and issues that arise within the program. Members of these panels and committees are distinguished as being among the most respected in their fields and provide WTR with the highest level of specialized input.

These committees and working groups are not scientific field-specific but content area-specific. For example, in the area of tumor promotion, the working group includes experts in fields ranging from toxicology to physics.

These committees and working groups are often asked to assist in peer review and to provide input on specific questions such as the feasibility of particular experimental assays for RF radiation exposure and the clinical significance of specific types of pacemaker interference.

Members of the Peer Review Board, government agencies, and industry have served on WTR working groups.

RISK EVALUATION WORKING GROUPS

Toxicology Working Group.

The Toxicology Working Group was involved in the development of the WTR

Research Agenda. Participants in this working group included the following:

Ian Munro, Ph.D.
Chair, Toxicology
Wireless Technology Research, LLC

Arthur W. Guy, Ph.D.
Chief, Dosimetry
Wireless Technology Research, LLC

David Brusick, Ph.D., A.T.S.
Scientific Director
Corning Hazelton, Inc.

Ernest E. McConnell, D.V.M, M.S.
Consulting Pathologist

Donald McRee, Ph.D.
Director of Extramural Research
Wireless Technology Research, LLC

R. Julian Preston, Ph.D.
Chemical Industry Institute of Toxicology

Thomas J. Slaga, Ph.D.
AMC Cancer Research Center

Andrew Sivak, Ph.D., F.A.T.S.
Consultant
Environmental Health Sciences

Gary Williams, M.D.
Chief, Division of Pathology and Toxicology
Director of Medical Sciences
American Health Foundation

Dosimetry Working Group.

The Dosimetry Working Group was also involved in the development of the WTR Research Agenda and in the development of an in vivo exposure system. Participants in this working group included the following:

Arthur W. Guy, Ph.D.
Chief, Dosimetry
Wireless Technology Research, LLC

Quirino Balzano, Ph.D.
Director
Florida Corporate Electromagnetic Research Laboratory
Motorola, Inc.

Chung-Kwang Chou, Ph.D.
City of Hope National Medical Center

David Dini, P.E.
Research Engineer
Underwriters Laboratories, Inc.

Camelia Gabriel, Ph.D.
Managing Director
Microwave Consultants, Ltd.

Om Gandhi, Sc.D.
Professor and Chairman
Department of Electrical Engineering
University of Utah

Neils Kuster, Ph.D.
Professor
Laboratory for EMF and Microwave Electronics
ETH Zurich

Ronald Petersen
Non-ionizing Radiation Manager
Radiation Protection Department
Lucent Technologies, Inc.

Peter Polson, Ph.D.
Principal
AUSA Research

R. Julian Preston, Ph.D.
Chemical Industry Institute of Toxicology

Allen Taflove, Ph.D.
Professor, Department of Electrical Engineering and Computer Science
Northwestern University

Tumor Promotion Expert Panel.

The Tumor Promotion Panel was charged with determining whether there were reasons to consider tumor promotion as a possible mechanism for RF effects. Participants in this working group included the following:

Larry E. Anderson, Ph.D.
Bioelectromagnetics Program Manager
Battelle Pacific Northwest Laboratories

Samuel Cohen, Ph.D.
Department of Pathology and Microbiology
University of Nebraska Medical Center

Ronald Petersen
Non-ionizing Radiation Manager
Radiation Protection Department
Lucent Technologies, Inc.

Henry C. Pitot, Ph.D.
McArdle Laboratory for Cancer Research
University of Wisconsin

Thomas J. Slaga, Ph.D.
AMC Cancer Research Center

Andrew Sivak, Ph.D., F.A.T.S.
Environmental Health Sciences

DNA/Genotoxicity Expert Panel.

The DNA/Genotoxicity Expert Panel was charged with evaluating and making recommendations on mechanistic studies and genotoxicity studies designed to assess the potential effects of RF on DNA. Participants in this working group included the following:

Richard Albertini, Ph.D.
Vermont Cancer Center
University of Vermont

David Brusick, Ph.D., A.T.S.
Scientific Director
Corning Hazelton, Inc.

Phillip Hanawalt, Ph.D.
Herrin Biology Laboratories
Stanford University

Ronald Petersen
Lucent Technologies, Inc.

R. Julian Preston, Ph.D.
Chemical Industry Institute of Toxicology

Donald McRee, Ph.D.
Director of Extramural Research
Wireless Technology Research, LLC

Gary Williams, M.D.
Chief, Division of Pathology and Toxicology
Director of Medical Sciences
American Health Foundation

Single Cell Gel Assay Expert Panel.

The Single Cell Gel Assay Expert Panel prepared a position paper on the single cell gel (SCG) assay and its strengths, limitations, and usefulness as a test for genotoxicity. Participants in this working group included the following:

Peggy Olive, Ph.D.
Medical Biophysics Unit
British Columbia Cancer Research Centre

R. Julian Preston, Ph.D.
Chemical Industry Institute of Toxicology

Richard Setlow, Ph.D.
Chair, Biology Department
Brookhaven National Laboratories

Raymond Tice, Ph.D.
Vice President, Research and Development
Integrated Laboratory Systems

Epidemiology Working Group.

The Epidemiology Working Group was involved in the development of the WTR Research Agenda. Participants in this working group included the following:

George L. Carlo, Ph.D., M.S., J.D.
Chairman and Chief, Epidemiology
Wireless Technology Research, LLC

Patricia Buffler, Ph.D., M.P.H.
Dean, School of Public Health
University of California at Berkeley

Nancy Dreyer, Ph.D.
President
Epidemiology Resources, Inc.

Saxon Graham, Ph.D.
Professor Emeritus
State University of New York at Buffalo

Kenneth Rothman, Dr.P.H.
Senior Epidemiologist
Epidemiology Resources, Inc.

Rebecca A. Steffens, M.P.H.
Staff Scientist, Epidemiology
Wireless Technology Research, LLC

Exposure Assessment Working Group.

The Exposure Assessment Working Group studied a number of cellular technology and human exposure issues in order to provide guidance in epidemiological studies. Participants in this working group included the following:

George L. Carlo, Ph.D., M.S., J.D.
Chairman and Chief, Epidemiology
Wireless Technology Research, LLC

Chung-Kwang Chou, Ph.D.
City of Hope National Medical Center

Quirino Balzano, Ph.D.
Director
Florida Corporate Electromagnetic Research Laboratory
Motorola, Inc.

Donnie P. Funch
Epidemiology Resources, Inc.

Arthur W. Guy, Ph.D.
Chief, Dosimetry
Wireless Technology Research, LLC

Jack Mandel, Ph.D., M.P.H.
Mayo Chair in Public Health
University of Minnesota

Susan Preston-Martin, Ph.D.
Professor
Department of Preventive Medicine
University of Southern California

Robert Morgan, M.D., S.M.Hyg.
President
Environmental Health Strategies, Inc.

Kenneth Rothman, Dr.P.H.
Senior Epidemiologist
Epidemiology Resources, Inc.

Rebecca A. Steffens, M.P.H.
Staff Scientist, Epidemiology
Wireless Technology Research, LLC

Tissue Type Working Group.

The Tissue Type Working Group studied five distinct tissue types of the head, neck, and upper body that may be affected by exposure to low-power devices such as handheld cellular phones. Participants in this working group included the following:

George L. Carlo, Ph.D., M.S., J.D.
Chairman and Chief, Epidemiology
Wireless Technology Research, LLC

Arthur W. Guy, Ph.D.
Chief, Dosimetry
Wireless Technology Research, LLC

Susan Preston-Martin, Ph.D.
Professor
Department of Preventive Medicine
University of Southern California

Kenneth Rothman Dr.P.H.
Senior Epidemiologist
Epidemiology Resources, Inc.

David Savitz, Ph.D.
Professor, Department of Epidemiology
School of Public Health
University of North Carolina at Chapel Hill

Rebecca A. Steffens, M.P.H.
Staff Scientist, Epidemiology
Wireless Technology Research, LLC

Clinical Significance Working Group.

The Clinical Significance Working Group discussed and defined the clinical significance of various forms of electromagnetic interference as they relate to pacemaker patients. Participants in this working group included the following:

Alan D. Bernstein, Eng.Sc.D., F.A.C.C.
Director of Technical Research
Newark Beth Israel Medical Center

Jay Gross, M.D.
Director, Intercampus Pacemaker Service
Montefiore Medical Center

David L. Hayes, M.D.
Director of Pacing Laboratory
Department of Cardiology
Mayo Clinic

Marleen E. Irwin, R.C.R.T.
Clinical Research Support Specialist
Grey Nuns Community Health Center

M. Osborn
Department of Cardiology
Mayo Clinic

Paul J. Wang, M.D.
New England Medical Center

RISK MANAGEMENT WORKING GROUPS

Product Certification Working Group.

The Product Certification Working Group was charged with developing standardized testing techniques for specific absorption rate (SAR) certification of wireless instruments. Participants in this working group included the following:

Quirino Balzano, Ph.D.
Director
Florida Corporate Electromagnetic Research Laboratory
Motorola, Inc.

Chung-Kwang Chou, Ph.D
City of Hope National Medical Center

David Dini, P.E.
Research Engineer
Underwriters Laboratories, Inc.

Om Gandhi, Sc.D.
Chairman, Department of Electrical Engineering
University of Utah

Arthur W. Guy, Ph.D.
Chief, Dosimetry
Wireless Technology Research, LLC

Wolf Pavlok
Corporate Vice President and General Manager
Motorola, Inc.

Ronald Petersen
Non-ionizing Radiation Manager
Radiation Protection Department
Lucent Technologies, Inc.

Product Design Changes Working Group.

The Product Design Changes Working Group was charged with consideration of potential system, portable, and accessory design options to lower or otherwise modify RF exposure. Participants in this working group included the following:

Wolf Pavlok
Corporate Vice President and General Manager
Motorola, Inc.

Jesse Russell
Vice President, Advanced Communications Technologies

AT&T

Ericsson

GE

MCC/Panasonic

Mitsubishi

Motorola

NEC

Nokia

Northern Telecom

Oki

Sony

Toshiba

Infrastructure Modifications Working Group.

The Infrastructure Modifications Working Group was charged with both evaluating the types of feasible infrastructure changes to modify power (and therefore RF exposure) from wireless instruments and assessing any risks associated with infrastructure modifications. Participants in this working group included the following:

George L. Carlo, Ph.D., M.S., J.D.
Chairman and Chief, Epidemiology
Wireless Technology Research, LLC

Candy Castle
Director, External Affairs
AT&T Wireless Services, Inc.

Arthur W. Guy, Ph.D.
Chief, Dosimetry
Wireless Technology Research, LLC

Keith Paglusch
Vice President, Engineering and Network Operations
Sprint Cellular

Ronald Peterson
Non-ionizing Radiation Manager
Radiation Protection Department
Lucent Technologies, Inc.

Michael Roden
Director, Network Operations
Southwestern Bell Mobile Systems

Richard Tell, Ph.D.
President
Richard Tell Associates, Inc.

James Tozzi, Ph.D.
Director
Federal Focus, Inc.

Usage Restrictions Working Group.

The Usage Restrictions Working Group was charged with evaluating emergent scientific information to make usage recommendations that would serve to mitigate identified public health risks. Participants in this working group included the following:

George L. Carlo, Ph.D., M.S., J.D.
Chairman and Chief, Epidemiology
Wireless Technology Research, LLC

Chung-Kwang Chou, Ph.D.
City of Hope National Medical Center

Arthur W. Guy, Ph.D.
Chief, Dosimetry
Wireless Technology Research, LLC

Kenneth Rothman, Dr.P.H.
Senior Epidemiologist
Epidemiology Resources, Inc.

Screening Working Group.

The Screening Working Group compared the results of bench testing and clinical testing for interference among cardiac pacemakers and cellular telephones and assessed how well the bench testing might predict this interference in the clinical setting. Participants in this working group included the following:

George L. Carlo, Ph.D., M.S., J.D.
Chairman and Chief, Epidemiology
Wireless Technology Research, LLC

Ronald Kaczmarek, M.D., M.P.H.
Medical Officer, Office of Surveillance and Biometrics
Center for Devices and Radiological Health
Food and Drug Administration

Paul Ruggera, P.E.
Special Projects Engineer, Electrophysics Branch
Center for Devices and Radiological Health
Food and Drug Administration

Mel Seidman
Division of Biostatistics
Center for Devices and Radiological Health
Food and Drug Administration

Rebecca A. Steffens, M.P.H.
Staff Scientist, Epidemiology
Wireless Technology Research, LLC

Claudine M. Valmonte, M.A.
Staff Scientist, Clinical Studies
Wireless Technology Research, LLC

Donald Witters, M.S.
Physicist, Electrophysics Branch
Center for Devices and Radiological Health
Food and Drug Administration

SCIENTIFIC OUTREACH WORKING GROUP

Education and Scientific Outreach Working Group.

The Education and Scientific Outreach Working Group was charged with developing programs and informative materials to educate the public regarding all aspects of the SAG research program. Participants in this working group included the following:

George L. Carlo, Ph.D., M.S., J.D.
Chairman and Chief, Epidemiology
Wireless Technology Research, LLC

Mary Beth Franklin
Franklin News Features

Jeffrey Nesbit
Director of Public Affairs
Wireless Technology Research, LLC

Michael Volpe
President
Volpe Communications

Penny Pagano
Freelance Writer

PACEMAKER INTERFERENCE EXAMPLE

This program began in June 1994 following the identification of a possible public health risk through initial reports of electromagnetic interference (EMI) between wireless telephones used in Europe and Australia and cardiac pacemakers. As a result, WTR, with the support of the wireless communication industry and pacemaker manufacturers, and in coordination with the FDA, took the lead in the multidisciplinary evaluation by conducting a multi-center clinical study and a risk management clinical study and by convening several expert working groups. This 27-month process resulted in a comprehensive database on which recommendations for corrective intervention could be made.

The goals of the WTR research program were to understand the public health impact of the interference between wireless phones and pacemakers and to make informed risk management recommendations for corrective intervention as needed. To meet these goals, WTR completed a comprehensive evaluation of all relevant data. These data were integrated into the risk management decision-making framework, and based on an evaluation of this full spectrum of data, WTR developed a series of comprehensive risk management recommendations.

After an evaluation of the science, WTR conducted a multi-center clinical study based on a three-part process. In the first part, the protocol for the clinical study was developed. A multidisciplinary protocol development committee coordinated by WTR was convened. This committee included representatives from the FDA, the wireless phone and medical instrument industries, the Center for the Study of Wireless Electromagnetic Compatibility, WTR, and clinicians from the Mayo Clinic, The George Washington University Medical Center, and Mount Sinai Medical Center. The committee held numerous meetings, conducted pilot tests, and edited protocol drafts in an effort to gather information and complete the collaborative development of the protocol. The second part entailed the implementation of the multi-center clinical study led by Dr. David Hayes at the Mayo Clinic in Rochester, Minnesota; Dr. Paul Wang and Dr. N.A. Mark Estes III at the New England Medical Center in Boston, Massachusetts; and Dr. Dwight Reynolds at the University of Oklahoma Health Sciences Center in Oklahoma City, Oklahoma. In the spring of 1995, after review by WTR's Peer Review Board, WTR approved the protocol for testing. Clinical testing was initiated in August of 1995 and terminated on February 29, 1996, with 980 patients participating in the study. The third part of the program included data analysis, data interpretation, and reporting of the results of the clinical study, including special attention to the clinical significance of findings.

WTR placed considerable importance on clinical significance classification. Clinical significance in this case is the health impact of EMI on a patient and the key to understanding the public health risk associated with interference. In an effort to categorize EMI, a framework in which to classify the events is desirable. To establish this framework and develop the important concept of clinical significance of EMI, WTR convened a panel of experts in cardiac pacing. The panel developed a classification scheme for clinical significance which included three categories of interference: Class I, definitely clinically significant; Class II, probably clinically significant; and Class III, probably not clinically significant. This classification scheme was used to interpret the incidence of interference seen in the multi-center clinical study.

As the clinical trial passed the halfway point, WTR concluded that a risk of interference between wireless telephones and implanted pacemakers existed. Though extensive data were being gathered related to the prevalence of the interference and the combination of different phone technologies and pacemakers, WTR recognized a need for further information that could be used to assess the effects of various parameters on EMI between handheld wireless phones and implanted pacemakers. WTR developed a complementary study protocol looking at these parameters including the distance between the pacemaker and phone, the orientation of the phone, and the sensitivity setting of the pacemaker. This risk management study tested 100 patients and was conducted at Mount Sinai Medical Center in Miami, FL, by Dr. Roger Carrillo.

WTR also established a working group, in collaboration with the FDA, to address screening issues related to interference between wireless handheld phones and pacemakers. The purpose of the group was to compare the FDA's in vitro study findings of interference between wireless telephones and pacemakers with the WTR multi-center clinical study findings in an effort to assess if the present in vitro techniques would be an effective tool for screening for interference for the future. As a result, the screening working group established a methodology that utilizes statistical as well as epidemiological principles.

After a 27-month process and through the use of a comprehensive database, WTR formulated comprehensive risk management intervention recommendations. The comprehensive database included all data that were critical to making informed judgments regarding interference between wireless phones and pacemakers. By combining all of the areas of research, WTR established recommendations in two general categories: mitigation and research. The mitigation recommendations were action steps that could be initiated immediately to minimize the risk of interference. The research recommendations were future steps that WTR identified as significant to long-term management of the public health risk associated with EMI between wireless phones and pacemakers. WTR targeted these recommendations specifically to pacemaker wearers, physicians, wireless phone industry, pacemaker industry, and the pacemaker and wireless phone industries jointly.

The application of the public health paradigm to this problem was a success. At the end of the 27 months, WTR was able to identify the risk associated with interference between wireless phones and pacemakers, examine the extent of that risk, develop interventions to mitigate the risk, and develop strategies to implement those interventions. However, WTR did not conduct this process alone; it was a coordinated effort with the FDA, the wireless communication industry, and the pacemaker industry that allowed this program to succeed.

WTR'S SCIENTIFIC DECISION TREE FOR DEFINING PUBLIC HEALTH PROBLEMS

Information is received into the WTR decision-making framework as a function of either the WTR ongoing surveillance program or its research program. Positive findings in toxicological experiments or positive associations in epidemiological studies are evaluated in terms of public health impact.

The evaluative considerations for public health impact in toxicology and epidemiology are distinct. They involve both objective and subjective assessments of data and require specific expertise for evaluation. Peer review is critical. A positive finding in either a toxicological or a epidemiological study may lead to a public health intervention, more research, or continued surveillance, but there is no set formula defining what might be required in a given instance.

The key to making these evaluations in the WTR program is the attempt to make decisions about public health problems on the basis of what is known, rather than on the basis of what is unknown. Within that framework, for example, it is unlikely that a positive finding will rise to the level suggesting public health intervention without evidence of dose-response, biological plausibility, or concordance between toxicology and epidemiology. However, a positive finding in either toxicology or epidemiology, absent dose-response, may still lead to further research.

The evaluation of scientific data in the context of a public health threat and subsequent decisions about public health interventions can only be made on a case-by-case basis, after thoughtful scientific scrutiny and rigorous peer review. The WTR decision-making paradigm encompasses these principles.

WTR SCIENTIFIC OUTREACH

WTR has made scientific outreach an integral part of its program. Through presentations at professional meetings, publications in peer-reviewed journals, and sponsorship of symposia such as the 1995 State of the Science Colloquium with the International Coordinating Committee on Wireless Health Research, WTR is able to share its findings with the scientific community in a timely manner as well as become informed of emerging scientific developments worldwide.

WTR has an extensive bibliography of published work, as follows:

Pending

Carrillo RG, Embrey M, & Zebede J. Pacemaker interference and wireless telephones: A study to generate risk management recommendations. *PACE* (submitted).

Dreyer NA, Loughlin JE, & Rothman KJ. Cause-specific mortality in cellular telephone users. 1999 (submitted).

Muscat J. Cellular telephone use and risk of brain cancer. New England Journal of Medicine 1999 (submitted).

Muscat J. Cellular telephone use and risk of acoustic neuroma. New England Journal of Medicine 1999 (submitted).

Steffens RA, Valmonte CM, Carlo GL, Kaczmarek R, Ruggera PA, Witters DM, & Seidman M. A comparison of clinical and bench test results from pacemaker/wireless phone interference studies: Does bench testing adequately predict clinical pacemaker interference? *Journal of Clinical Engineering* (submitted).

1999

Carlo GL, editor. Alternative animal models in radio frequency radiation research. Washington, DC: Health Risk Management Group, Inc.; 1999.

Chou CK, Chan KW, McDougall JA, & Guy AW. Development of a rat head exposure system for simulating human exposure to RF fields from handheld wireless telephones. *Bioelectromagnetics* 1999; 20(suppl 4): 75-92.

Guy AW, Chou CK, & McDougall JA. A quarter century of in vitro research: A new look at exposure methods. *Bioelectromagnetics* 1999; 20(suppl 4): 21-39.

1998

Anderson LE. Extremely low frequency fields and cancer: Laboratory studies. *International Journal of Toxicology* 1998;17(supp3):47-57.

Brusick DJ, Albertini R, McRee D, Peterson D, Williams G, Hanawalt P, & Preston J. Genotoxicity of radiofrequency radiation. *Environmental and Molecular Mutagenesis* 1998;32:1-16.

Carlo GL, editor. *Wireless phones and health: Scientific progress.* Boston: Kluwer Academic Publishers; 1998. 415 p.

Cohen SM. Cell proliferation in the evaluation of carcinogenic risk and the inadequacies of the initiation-promotion model. *International Journal of Toxicology* 1998;17(supp3):129-142.

Petersen RC. Radiofrequency dosimetry: Wireless handsets. *International Journal of Toxicology* 1998;17(supp3):11-37.

Pitot HC. The progression of neoplasia, cell replication, and electromagnetic fields. *International Journal of Toxicology* 1998;17(supp3):59-108.

Sivak A. Critical review of tumor promotion and related experimental studies using RFR.

International Journal of Toxicology 1998;17(supp3):39-45.

Sivak A. Expert panel summary report, tumor promotion panel, Wireless Technology Research. *International Journal of Toxicology* 1998;17(supp3):3-9.

Slaga TJ. Tumor promotion and/or enhancement models. *International Journal of Toxicology* 1998;17(supp3):109-127.

1997

Carlo GL, Hayes DL, Johnson CM, & Steffens RA. Cardiac pacemakers and cellular phones. *The New England Journal of Medicine* 2 October 1997;337(14):1007.

Carlo GL, Hersemann SE, & O'Donnell SE. Wireless Technology Research, L.L.C.'s public health paradigm approach to assessing and managing health risks. *Human and Ecological Risk Assessment* February 1997;3(1):7-14.

Embrey M & Carlo GL. Wireless Technology Research, L.L.C.'s risk management research. *Human and Ecological Risk Assessment* February 1997;3(1):85-95.

Findlay GK, Carlo GL, & Johnson CM. The fruits of ongoing surveillance: Wireless Technology Research, L.L.C.'s clinical risk evaluation research. *Human and Ecological Risk Assessment* February 1997;3(1):75-84.

Guy, AW. The starting point: Wireless Technology Research, L.L.C.'s dosimetry risk evaluation research. *Human and Ecological Risk Assessment* February 1997;3(1):25-50.

Hayes DL, Wang PJ, Reynolds DW, Estes M III, Griffith JL, Steffens RA, Carlo GL, Findlay GK, & Johnson CM. Interference with cardiac pacemakers by cellular telephones. *The New England Journal of Medicine* 22 May 1997;336(21):1473-1479.

Safety assessment: Wireless Technology Research, L.L.C.'s toxicology risk evaluation research. *Human and Ecological Risk Assessment* February 1997;3(1):51-58.

Sivak A. One toxicology activity: Wireless Technology Research, L.L.C.'s tumor promotion panel. *Human and Ecological Risk Assessment* February 1997;3(1):59-66.

Steffens RA & Carlo GL. Postmarketing surveillance: Wireless Technology Research, L.L.C.'s epidemiology risk evaluation research. *Human and Ecological Risk Assessment* February 1997;3(1):67-73.

Sund KG & Carlo GL. Wireless Technology Research L.L.C.'s ongoing surveillance. *Human and Ecological Risk Assessment* February 1997;3(1):15-23.

Sund KG, O'Donnell SE, & Volpe MJ. Wireless Technology Research, L.L.C.'s outreach to the scientific community, government, industry, and others. *Human and Ecological Risk Assessment* February 1997;3(1):97-117.

1996

Carlo GL, Sund KG, Sebeny P, & O'Donnell S. Scientific dispute resolution: A new model employing public health principles. *Environmental Claims Journal* Winter 1995/1996;8(2):123-131.

Chou CK, Bassen H, Osepchuk J, Balzano Q, Peterson R, Meltz M, Cleveland R, Lin JC, & Heynick L. Radiofrequency electromagnetic exposure: Tutorial review on experimental dosimetry. *Bioelectromagnetics* 1996;17:195-208.

Funch DP, Rothman KJ, Loughlin JE, & Dreyer NA. Utility of telephone company records for epidemiologic studies of cellular telephones. *Epidemiology* May 1996;7(3):299-302.

Hayes DL, Carillo RG, Findlay GK, & Embrey M. State of the science: Pacemaker and defibrillator interference from wireless communication devices. *PACE* October 1996;19:1419-1430.

Rothman KJ, Chou CK, Morgan R, Balzano Q, Guy AW, Funch DP, Preston-Martin S, Mandel J, Steffens R, & Carlo GL. Assessment of cellular telephone and other radio frequency exposure for epidemiologic research. *Epidemiology* May 1996;7(3):291-298.

Rothman KJ, Loughlin JE, Funch DP, & Dreyer NA. Overall mortality of cellular telephone customers. *Epidemiology* May 1996;7(3):303-305.

WTR STAFF

George L. Carlo, Ph.D., M.S., J.D.
Chairman

Donald McRee, Ph.D.
Director of Extramural Research

Jeffrey Nesbit
Director of Public Affairs

Thorne Auchter
Government Relations

Kelly Sund, M.S.
Staff Scientist, General

Toxicology and Dosimetry

Arthur W. Guy, Ph.D.
Chief, Dosimetry

Ian Munro, Ph.D.
Chief, Toxicology

Graham Hook, Ph.D.
Staff Scientist, Toxicology

Earle Nestmann, Ph.D.
Staff Scientist, Toxicology

Marjan Najafi, M.P.H.
Staff Scientist, Toxicology and Dosimetry

Kathleen Kapetanovic, M.P.H., J.D.
Staff Scientist, Toxicology and Dosimetry

Martha Embrey, M.P.H.
Staff Scientist, Toxicology

Susan Hersemann
Staff Scientist, Dosimetry

Epidemiology

George L. Carlo, Ph.D., M.S., J.D.
Chief, Epidemiology

Rebecca A. Steffens, M.P.H.
Staff Scientist, Epidemiology

Clinical Studies

Claudine M. Valmonte, M.A.
Staff Scientist, Clinical Studies

Gretchen Findlay, M.P.H.
Staff Scientist, Clinical Studies

Scientific Support

Polly Thibodeau
Scientific Support

Courtney Kvancz
Scientific Support

Salem Fisseha
Scientific Support

Marybelle Ang
Scientific Support

Abla Mawudeku
Scientific Support

Elizabeth Adams, M.P.H.
Scientific Support

Sherry Farr
Scientific Support

Pete Sebeny
Scientific Support

Jennifer Cohen
Scientific Support

Sam Amirfar
Scientific Support

Scientific Outreach

Nancy Akers, M.P.H.
Scientific Outreach

Mary Supley
Scientific Outreach

Janis Lemke
Scientific Outreach

Sue O'Donnell
Scientific Outreach

Accounting

Bryan Eddins
Administrator

Robert Bailey
Accounting Support

Slavko Bradic
Accounting Support

Chapter Keyword Index

MIX
Papier aus verantwortungsvollen Quellen
Paper from responsible sources
FSC® C105338

If you have any concerns about our products,
you can contact us on
ProductSafety@springernature.com

In case Publisher is established outside the EU,
the EU authorized representative is:
Springer Nature Customer Service Center GmbH
Europaplatz 3, 69115 Heidelberg, Germany

Printed by Libri Plureos GmbH
in Hamburg, Germany